AF294324

Biologic System Evaluation with Ultrasound

James F. Greenleaf   Chandra M. Sehgal

Biologic System Evaluation with Ultrasound

With 59 Illustrations

Springer-Verlag
New York Berlin Heidelberg London Paris
Tokyo Hong Kong Barcelona Budapest

James F. Greenleaf, Ph.D
Department of Physiology and Biophysics
Mayo Clinic Foundation
Rochester, MN 55905, USA

Chandra M. Sehgal, Ph.D.
Associate Professor
Department of Radiology
University of Pennsylvania
Philadelphia, PA 19104, USA

Library of Congress Cataloging-in-Publication Data
Greenleaf, James F.
    Biologic system evaluation with ultrasound / James F. Greenleaf,
    Chandra M. Sehgal
        p.   cm.
    1. Ultrasonics in biology.   2. Ultrasonics in medicine.
    I. Sehgal, Chandra M.   II. Title.
        [DNLM: 1. Computer Graphics—congresses.   2. Computer Simulation—
    congresses.   3. Models, Structural—congresses.
    4. Ultrasonography—congresses.   QY 26.5 G814b]
    QP82.2.U37G74   1992
    616.07′543—dc20
    DNLM/DLC
    for Library of Congress   92-2266

Printed on acid-free paper.

Production managed by Christin R. Ciresi; manufacturing supervised by Jacqui Ashri.
Camera-ready copy provided by the authors.

9 8 7 6 5 4 3 2 1

ISBN-13:978-1-4613-9245-3   e-ISBN-13:978-1-4613-9243-9
DOI: 10.1007/978-1-4613-9243-9

Preface

This book is the outcome of a course that we taught at the 1988 Ultrasonics Symposium in Chicago, IL. The book is meant for engineers who are not acquainted with current methods of ultrasonic analysis of tissues. Some engineers already in the field of ultrasonics may find the book useful as a reference. The concept of associating the scattering hierarchy with a biological hierarchy was developed from work by Wagner and Insana et al. Although their concepts of scattering are more sophisticated than those in this book, the hierarchy idea lended itself well to unifying the scattering concepts used herein. The book begins with two chapters of introduction. The second chapter describes relationships between the biological and scattering hierarchies used later in the book. The third chapter describes scattering using graphical contexts in the spatial and Fourier domains. The fourth through sixth chapters describe Class 0 (absorption and speed), Class 1 (speckle), Class 2 (resolved point), Class 3 (specular), and Class 4 (motion) scattering and their association with the biological hierarchy (molecules, cells, tissues, organs, and function). The seventh chapter describes instruments used for biologic system evaluation and the eighth chapter describes computed tomographic methods of imaging. The book could not have been done without the greatly appreciated and extensive assistance given by my secretary, Ms. Elaine C. Quarve, and graphics artist, Ms. Christine A. Welch. Editing was done in the Editorial Department by Dr. Carol L. Kornblith and Ms. Mary K. Horsman. We hope this book contributes to a better understanding of biologic system evaluation with ultrasound.

J.F. Greenleaf and C.M. Sehgal

Contents

1
Scattering vs. Biologic Hierarchies

1.1 Introduction

The approach taken in this book toward ultrasonic scattering as a method of noninvasive evaluation of biologic systems is to divide the problem into separate scattering[1] classes. The concentration or size of scattering centers relative to the resolution cell[2] of the imaging system is used to develop a hierarchy of scattering classes that correlates with a hierarchy of biologic classes. Class 0 scattering occurs when there are only absorption or speed or perhaps nonlinearity variations present as scattering terms. Class 1 scattering occurs when the concentration of scatterers per resolution cell is high (25 or higher). This occurs in tissues such as blood or liver and results in speckle, the fine-grained noise familiar from laser light. Class 2 scattering occurs when the concentration of scatterers is less than about 1 per resolution cell. This can happen simultaneously with Class 1 scattering. Class 3 scattering occurs when the scatterers are large relative to the resolution cell and cause specular or mirror-like reflection. We include a final type of scattering, Class 4, which is associated with motion causing Doppler shifts in the returned signal.

This hierarchy of scattering can be associated directly with biologic components in the following way.

1. Class 0 scattering is associated with molecules as solutes in solvents such as water, the most common molecule in the body.
2. Class 1 scattering is associated with cells, depending on their concentration. Cells are the basic unit of life and are assembled in groups as tissues and organs throughout the body.
3. Class 2 scattering is associated with tissues in which the structural architecture (connective tissue) or other components such as lipids are scattered throughout the tissue in concentrations lower than 1 per resolution cell. These elements scatter independently and cause scattering distinguishable from speckle produced by the cell components themselves.
4. Class 3 scattering is caused by the borders of the organs and vessels and is often specular in nature.

---

[1] The term "scattering" describes the part of the ultrasonic pressure field that when added to the original field results in the measured field. Both the real (speed) and the imaginary (absorption) part of the refractive index can cause a scattered field [1, 2].

[2] The resolution cell is related to the size of the acoustic pulse used to probe the object. It has dimensions in the axial direction that depend on the bandwidth of the system and in the transverse direction that depend on the center frequency and the aperture of the system [3].

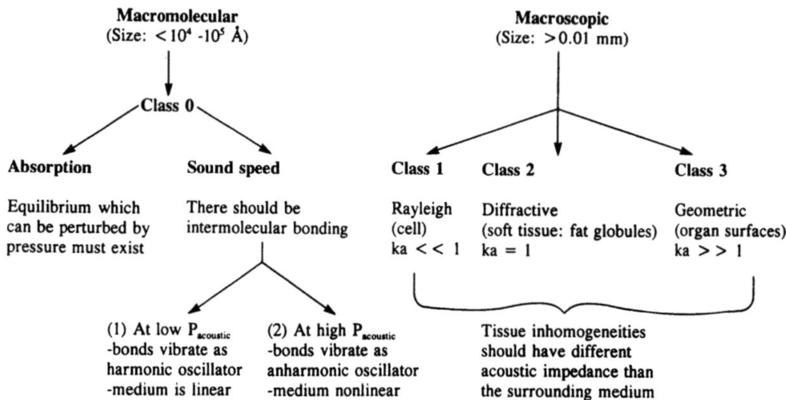

Figure 1.1. Depiction of scattering and biologic hierarchies used to describe ultrasound methods for evaluating biologic systems.

5. Class 4 scattering is caused by motion that produces a Doppler shift which is associated with ultrasonic signals scattering from interfaces within blood, heart, lung, and gut which move to accomplish their function.

1.2 Organization of Book

The book is formulated to describe the field of biologic system evaluation with ultrasound by using the scattering and biologic hierarchies described in Figure 1.1.

This book gives an overall view of ultrasonic imaging to the engineer or physicist who is unfamiliar with the field. It can also be used by the physician as a technical reference. We begin with a description of biologic systems, their hierarchic nature, their normal operation, and some of their pathologic features. Next, a very general description of scattering is used to explain the elements of scattering in a graphic context. Then, in three chapters, the scattering hierarchy is described in detail, with applications to biologic systems. Next, some typical backscattering imaging instruments are described with reference to the scattering classes that are used by the clinician to evaluate various pathologic conditions. Finally, some transmission imaging methods are described.

Bibliography

[1] J. F. Greenleaf, "An inverse view of scattering," *Proceedings IEEE Ultrasonics Symposium*, vol. 2, no. 2, pp. 821–824, 1984.

[2] A. J. Devaney, "Inverse source and scattering problems in ultrasonics," *IEEE Transactions on Sonics and Ultrasonics*, vol. SU-30, no. 6, pp. 355–364, 1983.

[3] R. F. Wagner, M. F. Insana, and D. G. Brown, "Unified approach to the detection and classification of speckle texture in diagnostic ultrasound," *Optical Engineering*, vol. 25, pp. 738–742, June, 1986.

2
Biologic Material Hierarchy

2.1 Introduction

The body is like an automaton which responds to inputs in such a way as to maintain its internal environment in a state required for survival. The fundamental element of the body is the cell. The body cells consist of a wide range of types, but all have common attributes such as requirements for energy and means of transport of metabolites and nutrients across their membranes. As far as ultrasonic imaging is concerned, the cells cause various amounts of scatter depending on their size, constituents, concentration, and hydration. Other components of the body of interest in this chapter are 1) molecules—proteins and lipids which when placed in solution with water, the most common molecule in the body, cause absorption of ultrasound in various degrees and variations in propagation speed; 2) tissues—aggregations of cells functioning for a relatively specific purpose; and 3) organs—aggregates of tissues dedicated to a general function.

The chapter is organized into a size hierarchy which parallels the ultrasonic scattering hierarchy described in Chapter 1. We proceed from molecules to cells to tissues to organs and finally to function (motion, for instance).

2.2 Molecules

The body consists of 60% water, 17% protein, 15% lipids, and the remainder of minerals, nucleic acids, and carbohydrate. Absorption of sound is related closely to the concentration of proteins in water. The proteins are very large molecules such as that depicted in Figure 2.1. Lipoproteins are a combination of lipids and proteins which have a more spherical shape (Fig. 2.2). The absorption of sound is related more to the concentration of the molecules than to their structure, as will be described in Chapter 4.

Liquid components of the body which most affect ultrasound are serum (the liquid environment of the blood cells), interstitial fluid (the environment of the cells of the remainder of the body), and intracellular fluid (the environment within the cells). Variations in concentrations of proteins and lipids affect ultrasound less than variations in amounts of fluids around the cells and within the tissues.

The extracellular fluid has a high concentration of chloride and sodium ions, much like the ocean from which the organisms emerged during evolution. Apparently the extracellular fluid is our "private ocean" which keeps the environment of the cells as it has been for millions of years. The intracellular fluid

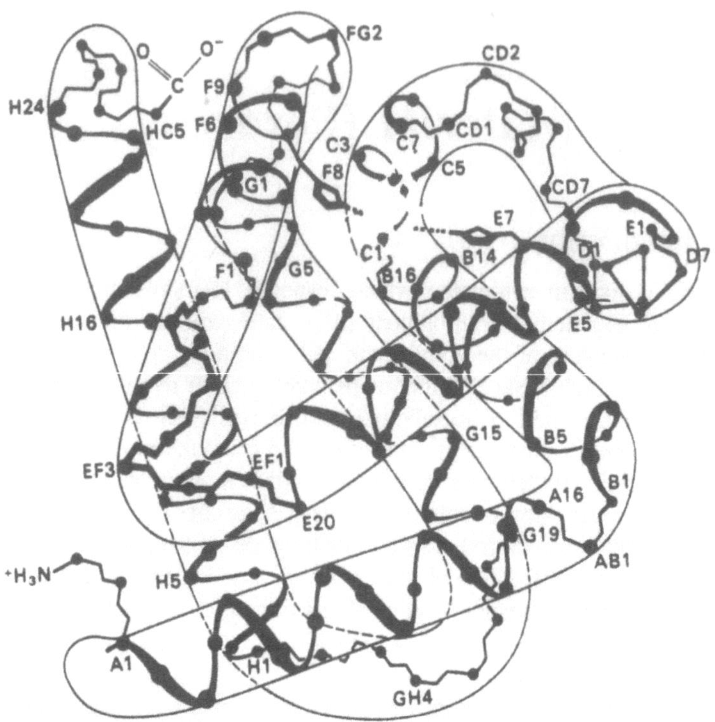

Figure 2.1. Proteins are complex and can have many forms, including the globular form shown here. (Reprinted with permission from Academic Press [1])

has a high concentration of potassium, a state which is maintained by ion pumps in the membrane of the cell that pump the sodium out and the potassium in.

The molecules of the extravascular fluid are maintained at certain concentrations for optimal functioning of the cell. The cells are so sensitive, however, that variations in concentration of the constituents of the fluids of the body that are inconsistent with life are often smaller than can be detected with noninvasive ultrasonic measurements.

The goal of the body as an integrated system is to maintain "homeostasis" or constancy in the internal environment of the individual living units—the cells. This is done with an elaborate organization of interrelated organ systems which control the molecular concentrations in the body fluids.

2.3 Cells

Cells are the basic living unit of the body. Their environment must be maintained so that they can carry out their function. Groups of similar cells are called tissues. Each organ is an aggregate of different combinations of tissues held

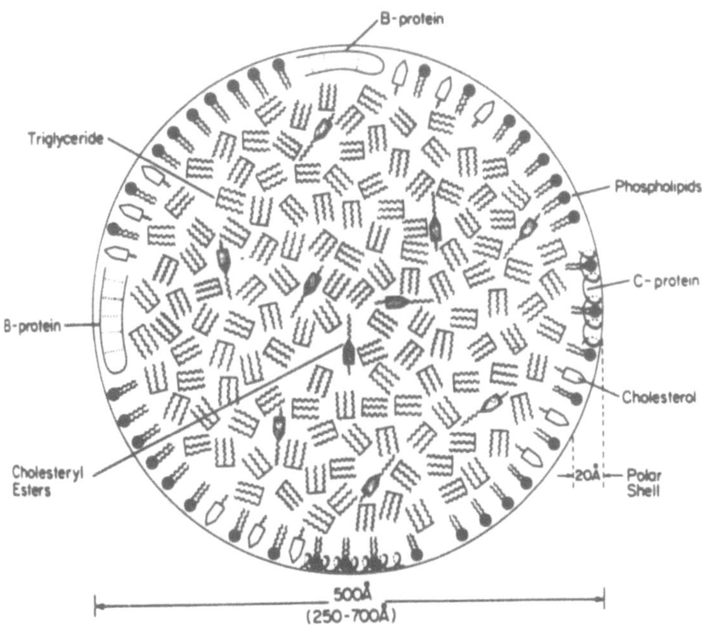

Figure 2.2. Lipoproteins are both lipid and protein and usually form spherical objects as shown. (Reprinted with permission from Elsevier Science Publishers BV [2])

together with intercellular supporting structures. Tissues and their component cells are adapted to do particular functions. In the tissue called blood the red blood cells, for instance, are oxygen transporters. There are about 75 trillion cells in the body.

All cells utilize almost identical types of nutrients. All cells use oxygen to produce energy by combining with carbohydrate, fat, or protein. All cells also deliver the end product of their activities to the extracellular fluid which then is excreted or transported to other parts of the body (Fig. 2.3).

Cells range in size from about 7 microns for a red blood cell to many centimeters for a nerve cell with its long dendrites. The cells are all bathed in an extravascular fluid which is in equilibrium with the blood serum flowing through the capillaries. Almost no cell is more than about 25 to 50 microns from a capillary. A notable exception, however, are the cells of the lens of the eye. As blood passes through the capillary a continual exchange of nutrients and metabolites occurs through the membranes of the capillary and the cell walls. The membranes of the cell wall can exchange large amounts of fluid very quickly.

The homeostatic conditions required by the body are provided by the organization of tissues into groups which each perform a function required to stabilize the internal environment. These groups of tissues are the organs of the body. The organs called the lungs, for instance, provide the oxygen input,

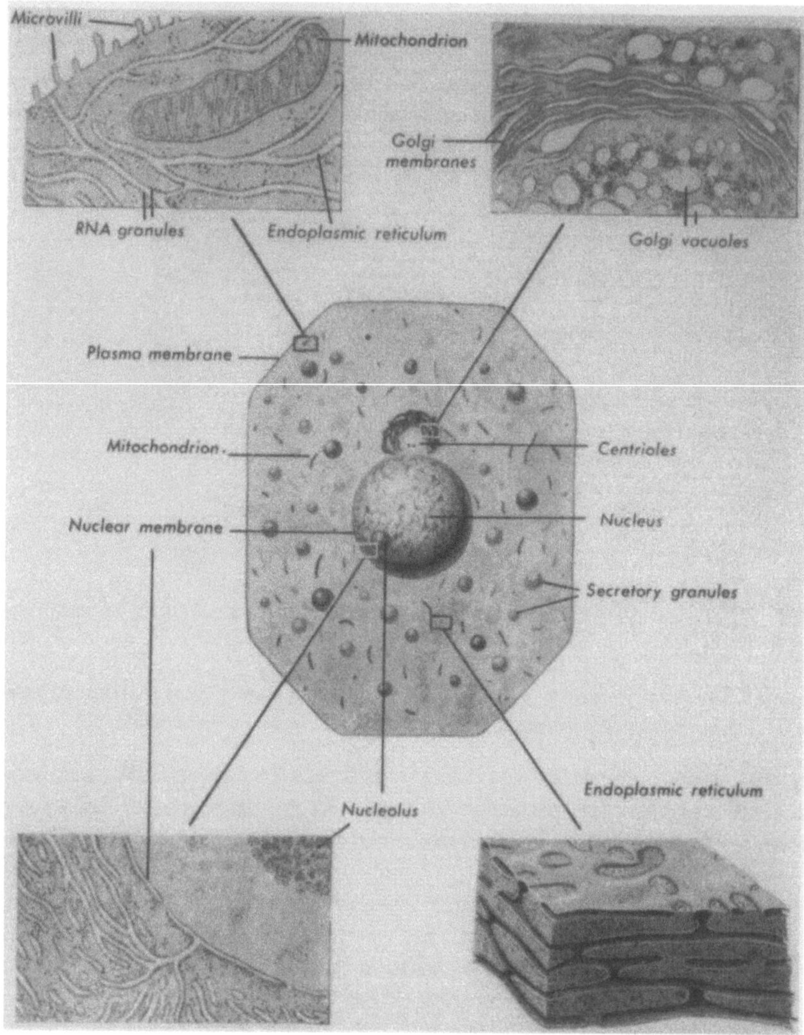

Figure 2.3. Schematic of typical cell and its components. The cell is the basic unit of life. (Reprinted with permission from The Williams and Wilkins Company [3])

whereas the kidneys stabilize the ion concentrations in the serum and, therefore, the extravascular fluid. The solid and hollow organs of the gastrointestinal system provide the nutrients and transport them into the serum for distribution throughout the body. Together, the tissues, organs, and systems produce the environment required for their own functioning and for the function of higher-order activities such as thinking, moving, and reproduction.

2.4 Tissues

Although ultrasonic scattering from the fluids of the body is confined to absorption or nonlinearity, cells cause absorption and scattering which are related in a very complex way to the frequency of ultrasound [4].

As stated previously, tissues are groups of cells with specific morphologic and physiologic characteristics. Tissues can be grouped generally into six types: 1) connective, 2) nerve, 3) muscle, 4) epithelial (lining), 5) hematopoietic, and 6) vascular. Blood, for instance, is a hematopoietic tissue composed of several types of cells such as the red cells and the white cells; physical transport is its main physiologic characteristic. The characteristics of tissues that are most important to ultrasound are the amount of structural proteins (collagen) and the concentration of the cells per resolution unit of the insonating ultrasound beam [4]. One of the most common processes that can influence these characteristics is inflammation.

2.4.1 Inflammation

Inflammation is a response of tissue to injury. It consists of a series of responses which have evolved to minimize the effects of injury. In response to injury, whether by trauma, heat, chemicals, or any other insult, the tissues release histamine and other substances that increase the permeability of the capillary membranes. This allows a large amount of protein and fluid to leak into the tissues. The result is edema or increased volume of interstitial fluid. This, in effect, decreases the concentration of cells within the tissue, something that can be detected with ultrasound as a change of backscatter or sound speed or attenuation (Fig. 2.4) [5]. Much of the fluid that flows into the injured area contains fibrin and may form a clot. This results in an effective "wall" around the injured area, blocking it from the remainder of the body. The inflammation may resolve and disappear over a period of weeks or it may develop into scar tissue which may be detected with ultrasound because of its high attenuation.

The alterations in backscattering ultrasound are not specific. However, if a highly specific question is asked such as whether there is inflammation or no inflammation, one could detect the injury very quickly, as has been shown previously [5].

After a few hours, the inflammation process continues with infiltration of the area with neutrophils (specialized white blood cells). This infiltration of cells once again increases the concentration of cells in the area and can cause a decrease in scattering as shown in Figure 2.4 in which the decrease of scattering was caused by a "mushy" cellular infiltration some 5 days after a heart transplant.

The final stage of inflammation consists of the movement of macrophages into the area of injury. Macrophages are cells that can increase in size greatly and phagocytose (engulf debris) some types of debris better than white cells can. The infiltration of macrophages along with large numbers of white cells often can result in a localized collection of dead liquefied tissue called pus. This collection of material is called an abcess. This response can often be

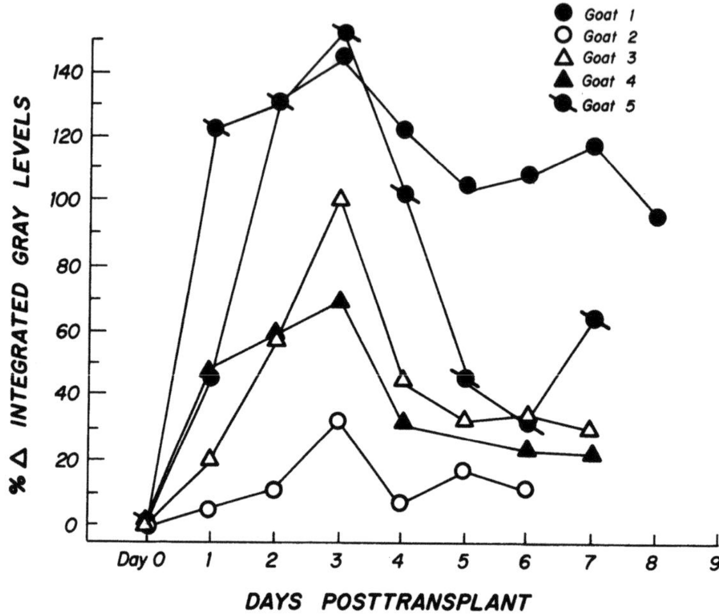

Figure 2.4. Average brightness of myocardium of animal heart transplanted into another animal as a function of time after the transplant. Tissue rejection occurs immediately. Brightness (backscatter) increases during the initial stages of inflammation (edema) and then decreases later during the macrophage infiltration stage of inflammation. (Reprinted with permission from International Society for Heart Transplantation [5])

detected with ultrasound by using backscatter imaging and results in dark areas in the image because the area has a high concentration of cells and is devoid of connective tissue so that it is much like a liquid and causes little backscatter at the frequencies used normally in medicine.

2.4.2 Fibrosis

Fibrosis occurs after inflammation as fibrin and other clotting mechanisms are followed by the appearance of cells called fibroblasts which synthesize collagen. A connective tissue scar is primarily composed of collagen. Fibrosis results in many "fibers" of connective tissue throughout the tissue. The long-term result is that injured areas often exhibit increased scattering and absorption of ultrasound after the resolution of the inflammation. Later stages of cirrhosis of the liver are often associated with fibrosis. The result is often an increased backscatter in liver images from such patients.

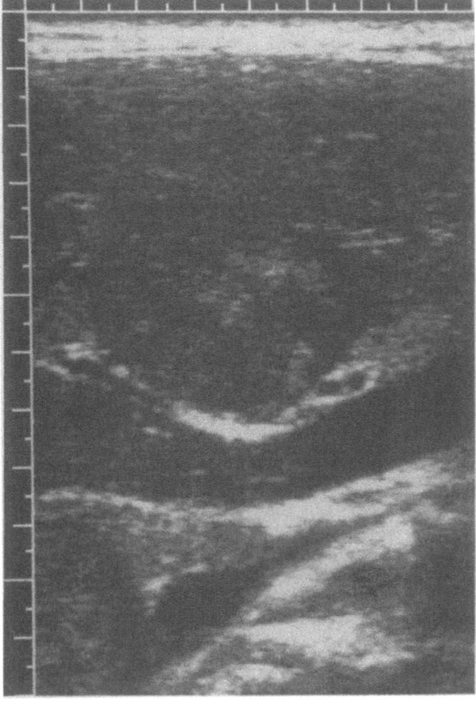

Figure 2.5. Ultrasonic image of a tumor in the liver which is displacing the portal vein.

2.4.3 Neoplasm

Neoplasm means new growth. It is used to refer to tumors. The word tumor is defined as a mass. A neoplastic tumor is an abnormal cell population with a capacity for progressive growth. The word oncology means the study of tumors. The cells of a tumor may be localized to a confined space such as solid tumors or they may be suspended in fluid such as in leukemia. There are two types of tumors, benign (simple or innocent) and malignant. Malignant tumors are often classified as carcinomas or sarcomas. Tumors composed of epithelial cells are called carcinomas. Those composed of connective, supportive, or muscle tissue are called sarcomas. Microscopically, tumors are composed of the cell population, the parenchyma, and the cells of the supportive and vascular tissue which is called the stroma [6]. Ultrasonic characteristics of benign or malignant tumors are similar because of the wide range of each type of tissue. Figure 2.5 is an image of a tumor in the liver which is displacing the portal vein. Tumors are often distinguished by their appearance relative to neighboring tissue or by their displacement of other tissue structures.

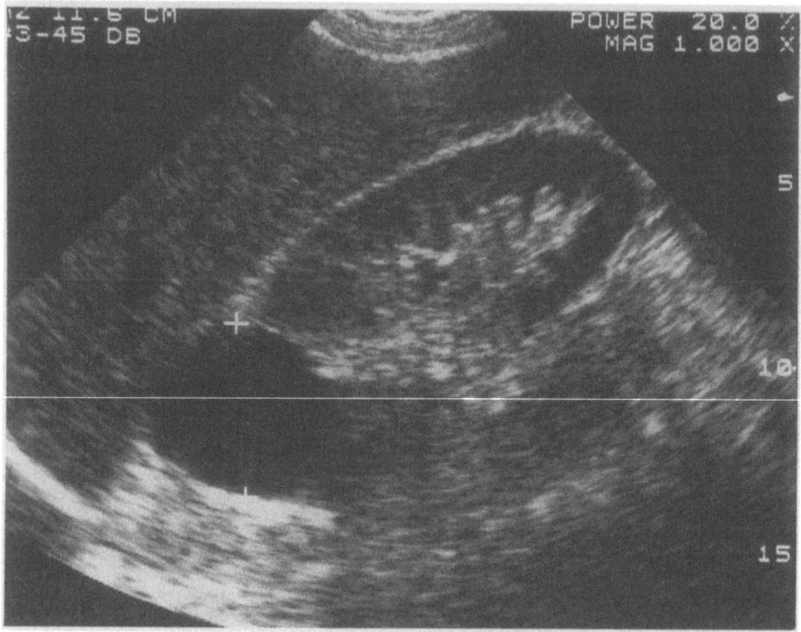

Figure 2.6. Typical image of kidney shows different types of cell scattering. A simple cyst is included in which no backscatter occurs within the fluid interior.

2.4.4 Other Diseases

There are a large number of diseases that alter the characteristics of the cell and the distribution of connective tissue. We have described here only a few of the more common conditions. For a review of medical ultrasound imaging see [7].

2.5 Organs

Some cells operate independently of other cells, such as the white cells of the blood, but more often they work as groups in tissues to perform as large functional units called organs, such as the heart, kidneys, lungs, and stomach. The kidney, for example, has a series of tubules lined with an endothelial layer of cells, nerve cells providing neural control, connective tissue, and cells holding it together and many types of cells making up the vessels, tubules, and associated smooth muscle layers. This range of cell and tissue types produces virtually all of the scatter classes (Fig. 2.6).

2.6 Function (Organ Systems)

Several tissues and organs function through motion, which can be detected either by real-time ultrasonic imaging or by Doppler detection of motion. Blood is a

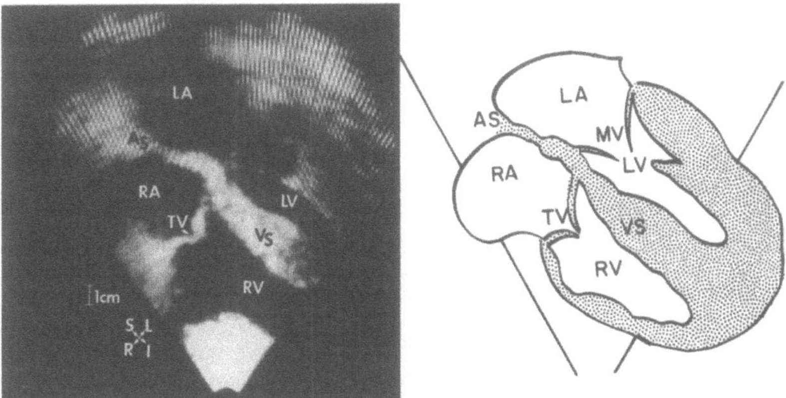

Figure 2.7. Image of the heart shows the four chambers and valves. Much information is gained from the dynamic analysis of these images (Class 4 scattering, motion, in addition to all other types of scattering). [AS, atrial septum; LA, left atrium; LV, left ventricle; MV, mitral valve; RA, right atrium; RV, right ventricle; TV, tricuspid valve; VS, ventricular septum.]

tissue that accomplishes its function of transport through motion and the heart is an organ that uses motion to accomplish its pumping function. Both the blood and the heart can be evaluated by real-time ultrasound. Other functions, which are not associated with motion, are more difficult to evaluate with ultrasound. For example, diseases of the liver or spleen are hard to detect with ultrasound, but alterations in the morphology of these organs can be used to deduce alterations in function. Similarly, the detection of tumors in the eye can be used to deduce that the eye is not functioning well but cannot evaluate the function of the eye. Several specific areas of imaging are described below and use the heart as an example. Many methods are in common use which evaluate pathologic states by using information about cells, tissues, and organs obtained noninvasively with ultrasound.

2.6.1 Heart

We will describe the heart in detail because it has specialized tissue and activity (motion) that accomplish its goal as a pump. In addition, the heart can be used as an example for describing a type of ultrasonic imaging that is highly successful. Ultrasonic imaging of the heart is used to evaluate its geometry as a function of time and the motion of blood within as well as its structural changes due to various pathologic states (Fig. 2.7).

The use of ultrasound in cardiology has revolutionized the field. Many of the routine measurements made in the catheterization laboratory are done now with ultrasound. Heart malformations, abnormal blood flow, and even clots are routinely detected with ultrasound [8]. To date, however, actual perfusion and muscle motion have not been evaluated with ultrasound.

The heart is a pulsatile, four-chamber pump. Two of the chambers, the atria, are entrances to the large chambers, the ventricles. One of the ventricles, the left, is a pressure pump producing high pressure blood flow to the arteries. The right ventricle is a volume pump which perfuses the lungs with blood at relatively low pressure.

The muscle tissue of the heart is much like striated muscle of the arms and legs but can beat more quickly for a much longer time (a lifetime). The nerve tissues of the heart are modified muscle fibers that conduct electrical stimuli throughout the heart to cause coordinated contraction of the heart and produce efficient pumping. Ultrasonic evaluation of the heart consists of images which contain reflections at the surfaces of the various borders (Fig. 2.7). These reflections are used to evaluate the motion of the various components of the heart. The backscatter from within the muscle is low (about 55 dB down from a perfect reflector) [9]. However, even this low level of Classes 2 and 3 scatter can be evaluated for some characteristics which can predict various pathologic conditions [5, 8]. The blood within the arteries of the heart and within the chambers of the heart produces Class 4 scattering by Doppler shift. A great deal of information about pathologic conditions of the heart can be determined from Doppler evaluations of the blood flow within the heart.

The period from the end of one heart contraction to the end of the next is called the cardiac cycle. Each cycle is initiated by a spontaneous excitation in a region between the atria and the ventricles called the AV node. This excitation continues throughout muscle of the heart guided by specialized bundles of conductors called the Purkinje fibers. Current ultrasonic imaging systems have frame rates of 30 per second, which is not fast enough to visualize the progress of contraction throughout the heart. Perhaps higher frame rates would be useful.

The valves between the atria and the ventricles are thin fibrous sheets that close when the pressure in the ventricles is greater than the low pressure in the atria. They are tethered by cords which are anchored in the ventricles. The input valves of the ventricles are thin fibrous bicuspid or tricuspid valves. The output valves are tricuspid semilunar valves that open and close passively depending on the differential pressure between the ventricles and the great vessels into which the ventricles pump. The valves can be evaluated with ultrasound. The size of their opening can be measured and the flow of blood across the valve can also be visualized by color Doppler flow imaging [10].

A few of the pathologic conditions that can be detected or evaluated with ultrasound are: the presence of clots in the chambers, pathologic congenital abnormalities in morphology, holes in the walls of the heart, stretched heart walls, abnormal wall thickness, abnormal wall motion, and leaky valves [10].

2.6.2 Other Organs

Most organs other than the heart do not move. Thus imaging with ultrasound requires the evaluation, either by eye or by computer, of the textures and patterns produced by the various scattering classes. The kidney shown in Figure 2.6

illustrates a wide range of scattering appearances. One of the best methods of evaluating not only vascular flow but tissue parenchyma is through color Doppler imaging [11]. This method encodes velocities measured through the Doppler shift of scattered ultrasound energy with color superimposed on the B-scan image. Flow in larger arteries and, perhaps, veins can be evaluated for abnormal characteristics by visual pattern recognition by the radiologist or subspecialist.

2.7 Summary

The body is made up of molecules in water (liquids), cells, tissues, organs, and organ systems. Each element of the biologic hierarchy has specific ultrasonic scattering characteristics. Ultrasound can be used to evaluate pathologic states by using its abilities to measure variations in each of the scattering modes, e.g., absorption in fluids (Class 0), altered scatter of cells (Classes 1 and 2), variations in geometry or morphology of organs (Class 3), and motion of surfaces of organs or of scatterers (Class 4). The fundamental mechanisms for scattering in each of these modes are developed in Chapters 4, 5, and 6.

Bibliography

[1] H. Neurath, *The Proteins: Composition, Structure, and Function*. 2nd Edition, volume 2. Academic Press, New York, 1964.

[2] J. D. Morrisett, R. L. Jackson, and A. M. Gotto, Jr., "Lipid-protein interactions in the plasma lipoproteins," *Biochimica et Biophysica Acta*, vol. 472, pp. 93–133, 1977.

[3] W. M. Copenhaver, *Bailey's Textbook of Histology*. The Williams and Wilkins Co., Baltimore, 1965.

[4] J. M. Thijssen, "Ultrasonic tissue characterization and echographic imaging," *Medical Progress Through Technology*, vol. 13, pp. 29–46, 1987.

[5] K. Chandrasekaran, R. C. Bansal, J. F. Greenleaf, A. Hauck, J. B. Seward, A. J. Tajik, and L. L. Bailey, "Early recognition of heart transplant rejection by backscatter analysis from serial 2D echos in a heterotopic transplant model," *Journal of Heart Transplantation*, vol. 6, pp. 1–7, 1987.

[6] D. M. Prescott and A. S. Flexer, *Cancer, The Misguided Cell*. Sunderland, MA: Sinauer Associates, First ed., 1982.

[7] P. N. Wells, "Blood flow: Insights from ultrasound," *Proceedings of the Institute for Mechanical Engineering [H]*, vol. 204, no. H1, pp. 1–20, 1990.

[8] M. O'Donnell, J. W. Mimbs, and J. G. Miller, "Relationship between collagen and ultrasonic backscatter in myocardial tissue," *Journal of the Acoustical Society of America*, vol. 69, pp. 580–588, 1981.

[9] J. G. Miller, J. E. Perez, and J. G. Mottley, "Myocardial tissue characterization: an approach based on quantitative backscatter and attenuation," *IEEE Transactions on Sonics and Ultrasonics*, vol. 32, p. 111, 1985.

[10] J. B. Seward, A. Tajik, D. Hagler, and W. D. Edwards, "Update on nomenclature, image orientation and anatomic-echocardiographic correlations with new tomographic views," in *Two-Dimensional Echocardiography* (J. N. Schapira, ed.), pp. 11–142, Williams and Wilkins. Baltimore, 1982.

[11] P. N. Burns, "The physical principles of Doppler and spectral analysis," *Journal of Clinical Ultrasound*, vol. 15, pp. 567–590, 1987.

3
Graphic Description of Scattering

3.1 Introduction

The *qualitative* goal of diagnostic imaging with ultrasound is to obtain images of tissue that are sensitive to disease processes.

The *quantitative* goal of diagnostic imaging with ultrasound is to relate metrics of the received signal to intrinsic physical tissue properties such as distributions of compressibility, density, and nonlinearity.

The current method of mathematically modeling the propagation of ultrasound through tissue is to simplify the viscoelastic wave equation to obtain a linear equation [1]. The Helmholtz equation represents a simplification of the full viscoelastic wave equation in which mechanisms such as shear, mode conversion, and flow have been eliminated [2].

There have been many attempts to relate intrinsic properties of elastic materials to measurable extrinsic properties such as pressure, temperature, or particle velocity of the scattered wave [3–6]. A goal of biomedical ultrasound imaging research is to estimate two- or three-dimensional distributions of intrinsic properties by insonating the tissue with a known ultrasonic wave and then relating the scattered wave to tissue properties through an appropriate model of wave propagation [7]. This process is often termed "inverse scattering" because the scatterer is determined from the scattering energy.

The currently accepted model for propagation of ultrasonic energy through tissue includes both refraction and diffraction mechanisms [8]. Therefore, unlike x-ray propagation, which can be assumed to travel in a straight line [9], propagation of ultrasonic energy is characterized by refraction and deflection. In addition, there are mechanisms of mode conversion, energy absorption, and stochastic scattering [8]. Current approaches to modeling the propagation of ultrasonic energy in tissue are based on the assumption that the Helmholtz equation is the governing wave equation. The simplified Helmholtz equation is converted to an inhomogeneous equation with constant coefficients. Several methods are used to transform the resulting wave equation into one that is tractable.

The first-order Born and Rytov transformations are usually used to produce a tractable wave equation [10, 11]. As early as 1969 Wolf realized small perturbation solutions existed for the inverse-scattering problem in the Born approximation. Since then, methods have been developed for solving the inverse-scattering problem in two spatial dimensions [12, 13], in two Fourier dimensions [14, 15], and in three dimensions [12, 16]. These approaches have been called "diffraction tomography" [14], although this is somewhat of a misnomer because the Born approximation is a true diffraction technique and the Rytov approximation includes not only diffraction but also refraction, as will be shown later.

All of the above solutions require simplification of the pertinent wave equations to an inhomogeneous Helmholtz equation with constant coefficients and a forcing function.

Recently, exact methods of solution have been developed for the Helmholtz equation [2, 17], but they are very difficult to implement and solve on a computer in a practical length of time.

This chapter describes a Fourier transform technique for calculating scattering, given the incident wave and material properties of the inhomogeneity (forward problem), and a similar Fourier method for estimating the inhomogeneities, given the incident wave and scattered field (inverse problem). The challenge of solving either the forward or the inverse problem is to obtain a computationally efficient algorithm that provides an interpretation having a minimum of aberrations due to the simplifications required to render the problem tractable. The following discussion will be limited to the Born and Rytov transformations [18], although some comments will be made about methods for obtaining higher-order approximations [1]. For a review of the field, see the special issue on digital acoustic imaging in IEEE Transactions on Sonics and Ultrasonics [19] or the special issue on inverse problems [20].

3.2 Wave Equation

We will consider the incident waves to be planar and of constant angular frequency. It is well known that other, more general, forms of insonation can be considered to be a superposition of planar, fixed-frequency waves in nonattenuating media [21]. The wave equation relates extrinsic variables (such as pressure) – those unrelated to the specific object under investigation – to the intrinsic properties unique to the object (such as density or compressibility). We will be working with small variations in pressure and therefore can assume the system to be linear – that is, doubling the incident pressure wave merely doubles the measured scattered pressure. For high pressure levels this is not true and nonlinear effects become important [22–24]. In addition, we will not treat the anisotropy of tissues [25]. A linear partial differential equation that seems to describe the wave propagation in isotropic soft tissues, both liquid and solid [26], is the inhomogeneous wave equation [21], p. 410.

$$[\nabla^2 + K^2]P(\hat{r},\omega) = -K^2\gamma_\kappa(\hat{r},\omega)P(\hat{r},\omega) + \nabla \cdot [\gamma_\rho(\hat{r},\omega)\nabla P(\hat{r},\omega)], \quad (3.1)$$

In this equation $K = \omega/c_0$ the wave number in the immersing liquid, ∇^2 is the Laplacian operator $(\delta^2/\delta x^2 + \delta^2/\delta y^2)$ in orthogonal Cartesian coordinates,

$$\gamma_\kappa = \frac{\kappa(\hat{r},\omega) - \kappa_0}{\kappa_0} \qquad (3.2)$$

and

$$\gamma_\rho = \frac{\rho(\hat{r},\omega) - \rho_0}{\rho_0} , \tag{3.3}$$

in which κ and ρ are the background compressibility and density and $c_0^2 = 1/(\kappa_0^2\rho_0^2)$.

The pressure distribution as a function of frequency $\omega = 2\pi f$ and position $\hat{r}$ is $P(\hat{r},\omega)$. The terms describing the object can be collected and designated by an operator Q, giving

$$Q(\hat{r},\omega) = -K^2\gamma_\kappa(\hat{r},\omega) + \nabla \cdot \gamma_\rho(\hat{r},\omega)\nabla. \tag{3.4}$$

This yields

$$[\nabla^2 + K^2]P(\hat{r},\omega) = QP(\hat{r},\omega). \tag{3.5}$$

In these equations, $P(\hat{r},\omega)$ represents the distribution of the extrinsic properties, such as pressure or particle velocity, whereas $Q(\hat{r},\omega)$ represents the distribution of material properties which are the forcing function or source term of Eq. (3.1). The forward problem is the procedure in which the distribution of P in domain Ω_0 (Fig. 3.1) is determined from the known probing waves ψ_0 and the known material properties (or induced source) Q. The inverse problem is the determination of Q within Ω from measurements of P within Ω_0. It must be noted that, if there are sources of ultrasonic energy within Ω, then, in general, the problem has no unique solution. In addition, the solutions of this problem are unique in two dimensions but not in three dimensions [27].

One can see from Eq. (3.1) that, although the left-hand side of the equation is a linear, spatially invariant, partial differential equation, the right-hand side includes the unknown distribution P within the entire domain Ω. This is a difficult situation to rectify because the distribution of P cannot be measured within Ω in order to solve for the unknowns γ_κ and γ_ρ. Therefore, Eq. (3.1) must be converted to what might be termed an "ideal equation" that can be solved efficiently [28]. To do this, the equation must be transformed; the most popular methods are the Rytov [11] and Born [10] transformations. Each of these results in an approximate solution in which terms other than the desired unknown (or image) Q appear. Discussions of the relative usefulness of the two approximations are given elsewhere [18, 29]. In general, it can be said that the Born approximation is best for scatterers that do not shift phase, such as small scatterers immersed in a constant background. The Rytov approximation is best when dealing with objects that shift the phase of the incoming wave, such as large, extended inhomogeneities.

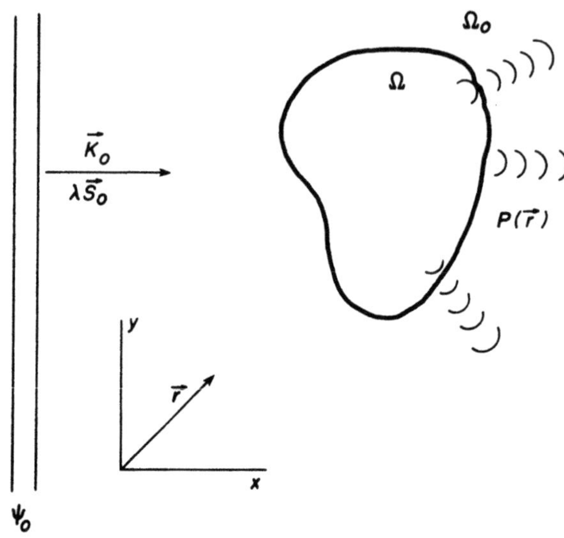

Figure 3.1. Definition of domain in which equations are defined. The support Ω contains the scatterer: Ω_0 contains only homogeneous material properties. $P(r)$ is the total pressure and $\vec{K}_0$ is the incident wave having direction $\vec{S}_0$ and wavelength λ. (Reprinted with permission from The Institute of Electrical and Electronics Engineers. Copyright 1984 IEEE [7])

3.3 Relationships Between the Fourier Transforms of P and Q

The solutions to be shown here are first-order solutions and are meant to give insight into the relationship between the space and frequency domains of the scattered waves and the scattering object. We derive the Born and Rytov approximations beginning with the Helmholtz equation and ending with an explicit equation in the form of Eq. (3.5) in which only the unknown Q is on the right-hand side. The Fourier relationships between the various functions will not be derived but will be described in a graphic context in an attempt to give insight into the first-order scattering relationships.

3.3.1 First-Order Approximations: Born and Rytov

Both the Born and Rytov transformations begin with Eq. (3.5). The two cases will be described separately.

The Born transformation assumes that the pressure P consists of the incident wave ψ_0 plus the scattered wave ψ_1.

Substituting $P = \psi_1 + \psi_0$ into Eq. (3.1) and assuming $\gamma_\rho = 0$ results in

$$\nabla^2\psi_1 + K^2\psi_1 = -K^2\psi_0\big(Q + \underline{Q\psi_1}\big), \tag{3.6}$$

in which $\psi_0 = exp(ikx)$, a plane wave (Fig. 3.1).

The Rytov transformation assumes a multiplicative relationship between the incident wave ψ_0 and the scattering wave $exp(\psi_s)$. Substituting $P = \psi_0 exp(\psi_s)$ into Eq. (3.5) and assuming $\gamma_\rho = 0$ gives

$$\nabla^2 \psi_2 + K^2 \psi_2 = Q\psi_0 + (\gamma_\rho \psi_0 |\nabla \psi_s|^2), \qquad (3.7)$$

in which $\psi_2 = \psi_0 \psi_s$ and the dependence on $\hat{r}$ and ω of the extrinsic variables ρ, ψ_1, ψ_2 and the intrinsic variable Q is implicit.

Note that the operators on the left sides of Eqs. (3.6) and (3.7) are identical; the right sides differ only by a second-order term (underlined in each case).

The second-order terms are the error terms because in the first-order Born or Rytov approximation they are dropped, giving an equation that can be solved in either the forward manner – i.e., for ψ_1 or ψ_2 or in the inverse manner for Q.

The solution of Eq. (3.1) in terms of the Fourier transforms of the quantities is as follows [27].

$$\Psi\left(\vec{K}, \vec{K}_0\right) = -K^2 \left[\Gamma_\kappa \left(\vec{K} - \vec{K}_0\right) + \vec{K} \cdot \vec{K}_0 \Gamma_\rho \left(\vec{K} - \vec{K}_0\right) \right], \qquad (3.8)$$

in which $\Psi(\vec{K}_0, \vec{K})$ is the angular spectrum of either ψ_1 or ψ_2 and $\Gamma_\kappa(\vec{K})$ and $\Gamma_\rho(\vec{K})$ are the spatial Fourier transforms of $\gamma_\kappa(\hat{r})$ and $\gamma_\rho(\hat{r})$. $\vec{K}$ is the wave vector in the direction of a scattered wave and $\vec{K}_0$ is the wave vector in the direction of the incident wave. The dependence of Γ_κ and Γ_ρ and thus Ψ on ω is implicit.

3.3.2 Fourier Relationships

We will begin describing the graphic scattering relationships with some simple diffraction gratings. The classical scattering of a plane wave having a direction unit vector $\vec{K}_0$ and wavelength λ impinging on a sinusoidal attenuation grating having a wavelength d is illustrated in Figure 3.2. The resulting scattered waves propagate in the direction of the zero order $\vec{S}_0$ and two first-order lobes $\vec{S}^1$ and $\vec{S}^2$, all of the original wavelength λ. The relationships between the scattering angle θ and d is $d sin\theta = \lambda$. This relationship can be constructed by using Huygen's wavelets in which one can see that the wave propagating in direction $\vec{S}^1$ is that wave in which Huygen's wavelets will constructively interfere after scattering from the grating. In addition to these transmitted waves, there are backscattered waves as well, although they are not shown in Figure 3.2.

Although simple for a grating, this classical evaluation of scattering is more complicated when the scattering object becomes complicated or when the angle of incidence is not orthogonal.

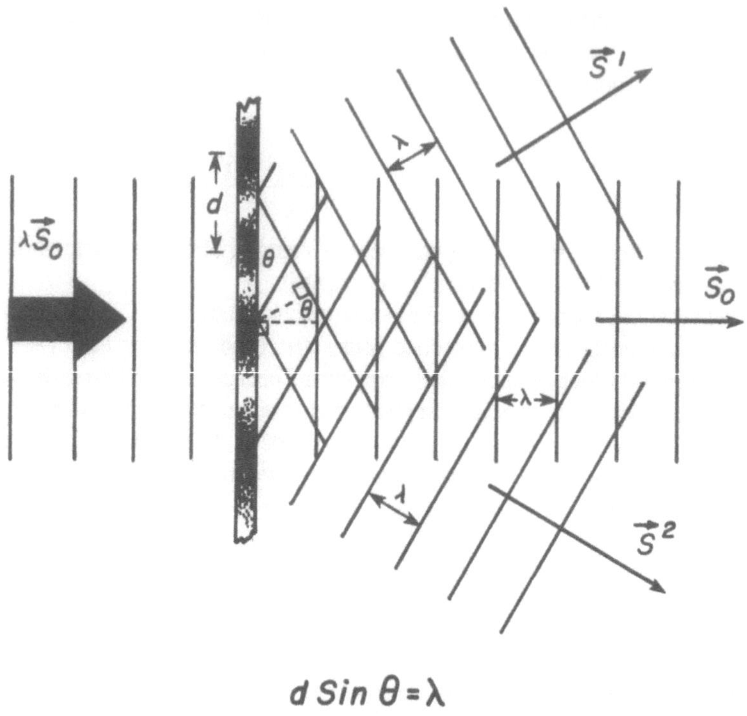

Figure 3.2. Classical scattering in which constructive interference results in scattered orders $\vec{S}^1$, $\vec{S}^2$, and $\vec{S}_0$. (Reprinted with permission from The Institute of Electrical and Electronics Engineers. Copyright 1984 IEEE [7])

3.3.3 Orthogonal Incidence

Figure 3.3 illustrates the relationship between the diffraction grating, the incident plane wave, and scattered waves in both the spatial domain and in the Fourier domain. Eq. (3.8) reveals that the angular spectrum of the scattered wave is proportional to the sum of two terms: 1) the Fourier transform of the fractional change in compressibility, κ evaluated at all scattering angles, $\vec{K}$, minus the incident wave direction, $\vec{K}_o$, plus 2) the scalar product of $\vec{K}$ and $\vec{K}_0$ multiplied by the density variation, ρ, evaluated at the same points as κ. A geometric method for determining the scattering in Fourier space can be used. We begin by drawing a circle of radius $|\vec{K}_0|$ centered at $-\vec{K}_0$. This circle represents all possible scattering from the object. That is, for each direction $\vec{K}_s$ indicated by a vector $\vec{K}_s = \vec{K} - \vec{K}_0$, there is a possible plane wave scattered from the object. However, the phase and magnitude of the scattered wave are defined by Eq. (3.8) and occur at points where the right-hand side is nonzero. Because the scattering object is a simple sinusoidal grating, we can derive its Fourier

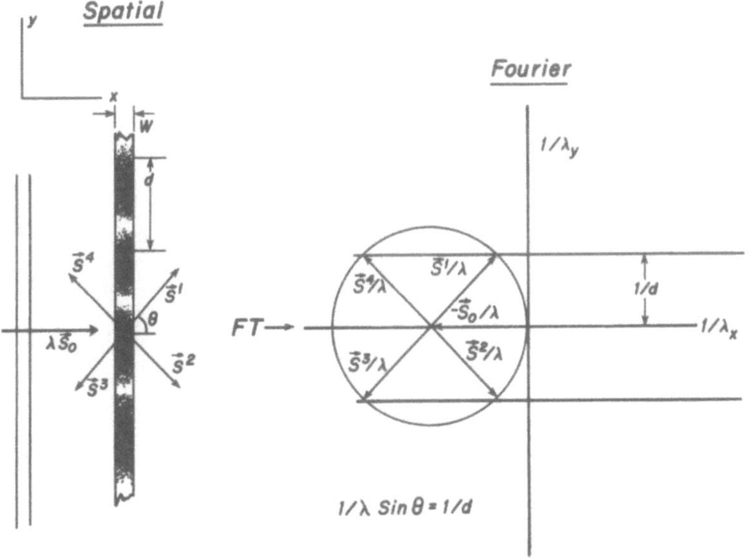

Figure 3.3. Relationship between spatial and Fourier domain depictions of the incident wave direction ($\vec{S}_0$) and scattered waves ($\vec{S}^x$) and the scatterer. Relationship is same as derived in Figure 3.2. (Reprinted with permission from The Institute of Electrical and Electronics Engineers. Copyright 1984 IEEE [7])

transform (the right side of Eq. [3.8]), and graph it on the Fourier plane in the following manner.

The equation for the diffraction grating is

$$\hat{Q}(x, y) = \gamma_\kappa \left[H\left(x + \frac{W}{2}\right) - H\left(x - \frac{W}{2}\right) \right] cos\frac{2\pi}{d}y, \qquad (3.9)$$

in which $H(x)$ is the Heaviside unit step function.

Its Fourier transform is

$$\hat{Q}(\mathrm{K}_x, \mathrm{K}_y) = \int\int \hat{Q}(x, y)e^{-i(\mathrm{K}_x x + \mathrm{K}_y y)}dx\,dy, \qquad (3.10)$$

giving

$$\hat{Q}(\mathrm{K}_x, \mathrm{K}_y) = \gamma_\kappa sinc(\mathrm{K}_x W)\left[\delta\left(\mathrm{K}_y - \frac{2\pi}{d}\right) + \delta\left(\mathrm{K}_y + \frac{2\pi}{d}\right)\right], \qquad (3.11)$$

shown in Figure 3.3.

The amplitude and phase of the Fourier transform of the grating depend on the thickness, w, so that the lines at $\pm 1/d$ have a sinc-like dependence

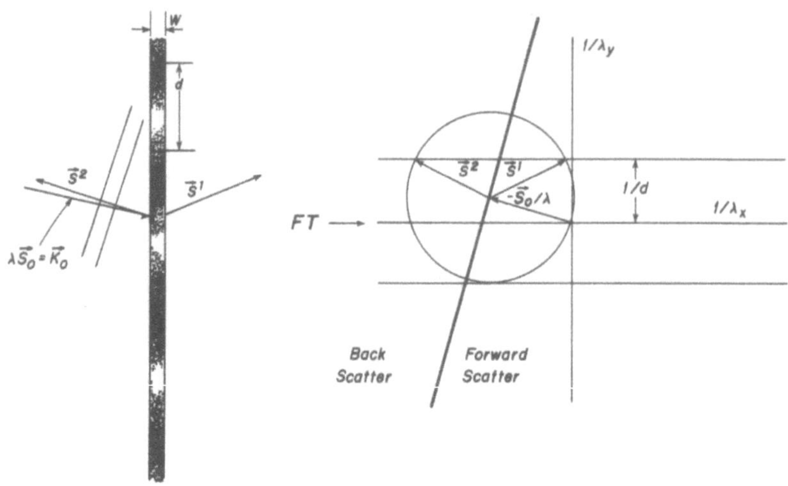

Figure 3.4. Same spatial and Fourier domain depiction as Fig.3.3 but for nonorthogonal incidence of $\vec{S}_0$. (Reprinted with permission from The Institute of Electrical and Electronics Engineers. Copyright 1984 IEEE [7])

in amplitude and phase for $|K_x| \rightarrow \infty$. The circle of possible scattering vectors having a radius $\left|\vec{K}\right|$ and centered at $\left|-\vec{K}_0\right|$ indicates that scattering can occur only at those points in the Fourier domain where the circle intersects nonzero values of the Fourier transform of the diffraction grating. This results in scattering possibilities in four separate scattering directions numbered $\vec{S}^1$ to $\vec{S}^4$ in Figure 3.3. In addition, depending on whether the value of the Fourier transform of the spatial grating is other than zero at the origin of the Fourier domain, there may be a zero-order wave propagating through the diffraction grating. Inspection of scattering wave S^1 reveals that the relationship between the scattering wave and the frequency of the diffraction grating is $\lambda sin\theta = 1/d$, the same relationship as derived from Figure 3.2 by using Huygen's constructions.

3.3.4 Nonorthogonal Incidence

The same geometric construction shown in Figure 3.3 is demonstrated in Figure 3.4 with nonorthogonal incidence on the grating. In this case the incident wave vector $-\vec{K}_0$ is drawn and a circle of radius $\left|\vec{K}\right|$ is again centered at $-\vec{K}_0$. This circle intersects the Fourier transform of the diffraction grating at two points representing scattered waves $\vec{S}^2$ and $\vec{S}^1$. It now becomes evident that scattering vectors in the direction of $\vec{K}_0$ are forward scattered and waves having vectors $\vec{K}$ pointing in the general direction of $\vec{K}_0$ are backscattered waves. This technique for constructing scattering waves from a given incident wave vector $\vec{K}_0$ and the Fourier transform of the complex compressibility or density of the scattered

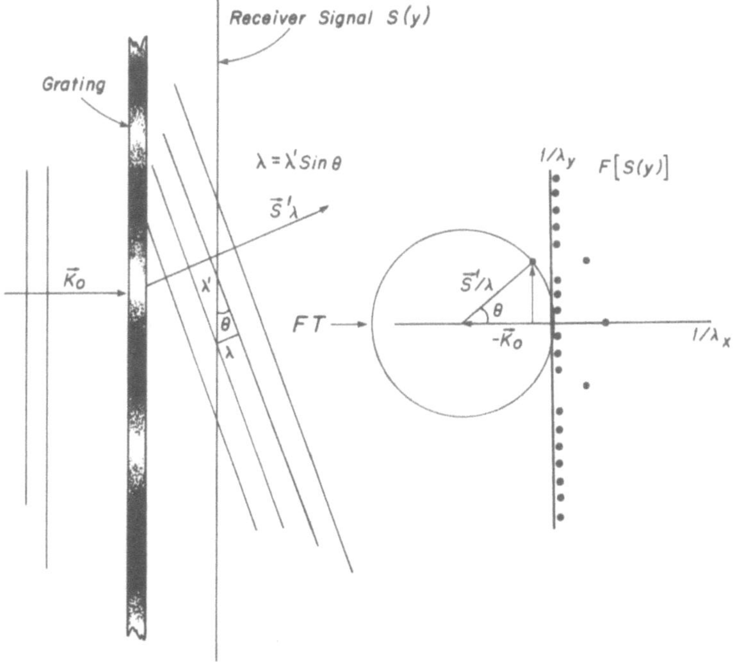

Figure 3.5. Method of determining the angular spectrum $\Psi(\vec{K}, \vec{K}_0)$ from a signal measured in the near field of the scatterer. The Fourier transform of the received signal is projected onto the circle centered at $-\vec{K}_0$. (Reprinted with permission from The Institute of Electrical and Electronics Engineers. Copyright 1984 IEEE [7])

object is a simple geometric method for constructing solutions to forward scatter problems and could be implemented on a computer. This procedure is known as "forward diffraction tomography" [14].

3.4 Inverse Diffraction Tomography

Inverse diffraction tomography is a method for obtaining estimates of the object scattering function from measurements of the scattered waves.

By rotating the incident wave vector $\vec{K}_0$ around the scattering object, multiple intersections of the related scattering circle with the Fourier transform of the object can be measured. Thus, measurements of the angular spectrum of scattering from many angles result in data required to fill in the Fourier domain using appropriate interpolation methods [28]. An estimate of the Fourier transform of the object can be produced from these data. This leaves us with the problem of measuring the angular spectrum $\Psi(\vec{K}, \vec{K}_0)$ of the scattered wave for each of these orientations $\vec{K}_0$.

The distribution of amplitudes and phases as a function of the direction vector $\vec{K}$ for each insonating direction $\vec{K}_0$ is called the angular spectrum of the scattered wave [27]. The angular spectrum of the scattered wave is not usually measured in the near field. However, a technique for determining the angular spectrum $\psi(\vec{K}, \vec{K}_0)$ from the total scattered wave in the near field is shown in Figure 3.5. The pressure wave is measured in the near field along a line locus orthogonal to $\vec{K}_0$ and results in a complex amplitude S(y). As seen in Figure 3.5, the angle θ of the scattered wave is related through a cosine law to the spatial wavelength of the wave as it intersects the receiver locus.

Therefore, the spatial amplitude and phase of the signal received along the line will vary with the wavelength as the hydrophone scans across the scattered wave having direction $\vec{S}^1$. A spatial Fourier transform of the amplitude and phase measured along the receiver locus will give a value at the frequency corresponding to $2\pi/\lambda'$ as shown in the Fourier domain of Figure 3.5. One can see that this frequency can be projected onto the scattering circle at the point representing the scattered wave $\vec{S}^1$. Therefore, projecting the values of the spatial Fourier transform of the measured signal onto the scattering circle gives the distribution of the angular spectrum around the circle. The same relationship holds in backscatter. Thus, the spatial Fourier transform of the signal, measured along a line in the near field that is orthogonal to the incident plane wave direction S_0, is related to the angular spectrum required in Eq. (3.8) through a cosine law which can be graphically constructed by using the projection shown in Figure 3.4.

3.4.1 Backscattering Analysis

Figure 3.6 illustrates the Fourier relationships for a backscattering experiment. One can see that the backscattering angles represented by one of the scattering waves at S are those angles directed toward the incident wave $\vec{K}_0$ in the spatial Fourier domain. Thus, by measuring backscatter from the object for many experiments with incident waves at many angles $\vec{K}_0$, the values of the Fourier transform of the object can be obtained within the annulus in Figure 3.6 described by

$$\sqrt{2}\left|\vec{K}_0\right| \leq \sqrt{K_x^2 + K_y^2} \leq 2\left|\vec{K}_0\right| . \tag{3.12}$$

This indicates that images of backscatter (B-scans), or at least compound B-scans (images taken from many angles and averaged), produce images of the higher-frequency region of the Fourier domain. Thus, the low-frequency (i.e., quantitative) region of the Fourier domain is not accessed by backscatter imaging. This is the reason that B-scan images appear to be somewhat differentiated or high-pass filtered. The differential operator in the Fourier transform domain is a ramp beginning at zero at the origin and increasing with radius from the origin, a function vaguely similar to the shaded annulus of Figure 3.6.

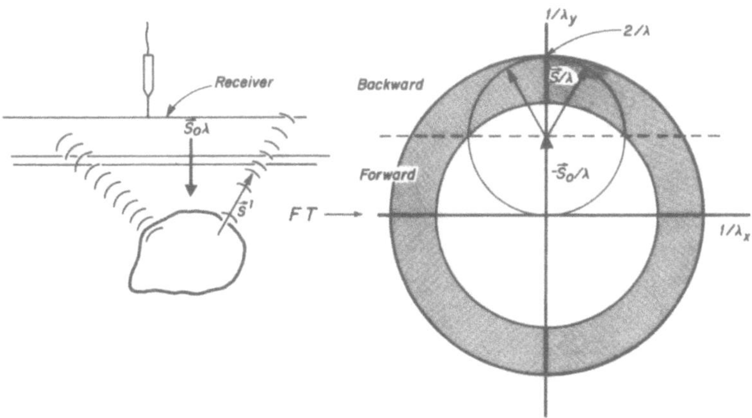

Figure 3.6. In backscatter tomography, only the shaded annulus can be calculated. This results in high-frequency estimates of the object but without the quantitative region of the Fourier domain (the interior disk). (Reprinted with permission from The Institute of Electrical and Electronics Engineers. Copyright 1984 IEEE [7])

3.4.2 Echography

The Fourier domain can be filled by either rotating the vector $\vec{K}_0$ or changing its length $|K_0|$. In typical B-scans, both are done. The rotation of $\vec{S}_0$ is done by focusing either with phased arrays or with fixed lenses; variation of the wavelength is obtained by transmitting a broadband pulse, thus using a multifrequency signal. Therefore, an approximation of the high-frequency region of the Fourier domain representation of the scatterer is obtained from single views or single passes of the transducers in the typical B-scan or phased-array image.

A B-scan or echogram is an image produced by a scanning apparatus and is a convolution of the point-source function (PSF) of the imaging system with the scattering function. If we let the PSF be $h(\hat{r})$ and the scattering function be $s(\hat{r})$, their spatial Fourier transforms are $H(\vec{K})$ and $S(\vec{K})$. The image $i(\hat{r})$ is a convolution between s and h. This is the same as the product of S and H in the Fourier domain as shown in Figure 3.7. The position of the PSF in the Fourier domain depends on the direction of the beam relative to the object. The bandwidth of the transducer gives the magnitude axial resolution $S_{ca}=1.37/\Delta f$, while the center frequency and the aperture size give the lateral resolution $S_{ce} = 0.87\lambda z_o/D$ where D is the diameter of the transducer, λ is the wavelength, and Δf is the bandwidth [30].

Therefore, the axial resolution is generally better than the lateral resolution. Only a small portion of the Fourier transform of the scattering function $S(\vec{K})$ is accessible with B-scan imaging.

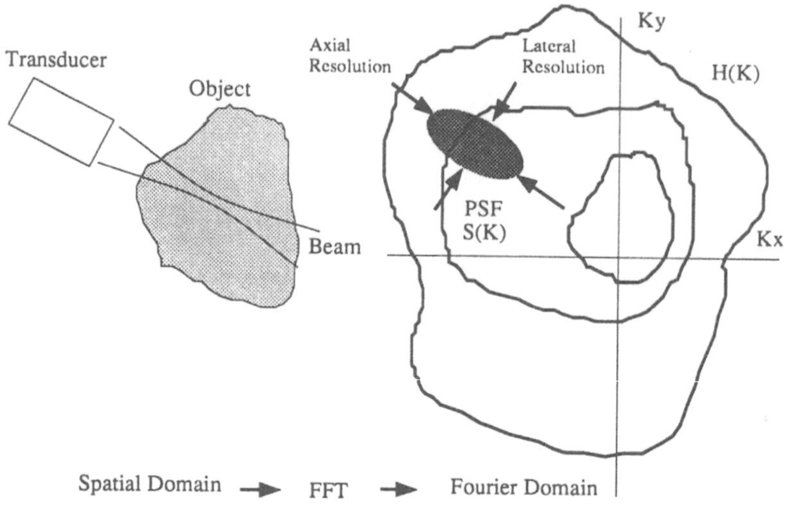

Figure 3.7. Schematic of the relationships between Fourier, spatial, and beam domains. The lateral resolution of the B-scanner depends on the range of angles of wave numbers insonating the object and the axial resolution depends on the range of frequencies (wavelengths) of the wave numbers insonating the object. PSF, point-source function.

3.5 Transmission Tomography

Figure 3.8 illustrates the relationships between the object and the scattered and incident waves in a transmission experiment. One can see that the scattered waves measured in the transmission tomography experiment are those waves directed toward the origin of the center of the angular spectrum circle; they are depicted by $\vec{S}^1$ and $\vec{S}^2$. Insonating from various angles results in a set of measurements in the interior circle of spatial Fourier domain having radius $\hat{R} \leq \sqrt{2}|K_o|$. Although the spatial frequencies available to transmission tomography experiments are lower than those available to backscatter tomography, transmission tomography results in quantitative images because the DC and low-frequency values of the Fourier transform of the object are measured. Multifrequency signals can be used in addition to multiple views [31].

A wide range of experiments can be performed in which various sectors of the angular spectrum circle are measured. A novel geometry was suggested by Nahamoo et al. [15] in which scattering from 90° is used to fill in the Fourier domain.

One can see from Eq. (3.8) that density scattering adds to the compressibility scattering but with a cosine dependence on the angle between the scattering direction and the incident beam. Mueller et al. [14] and others have noted that multifrequency experiments should provide a method of separating compressibility and density. Norton and Linzer [32] noticed this relationship and

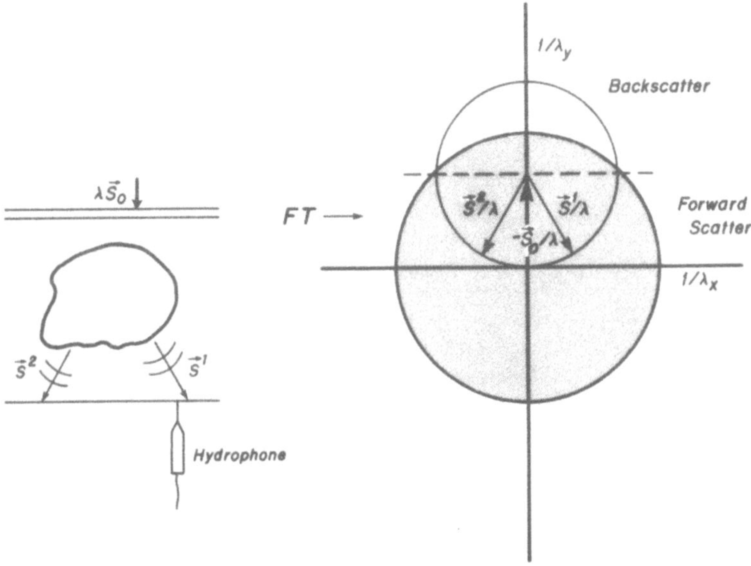

Figure 3.8. The region of the Fourier domain available to transmission tomography experiments is the shaded interior disk, resulting in low frequency but quantitative estimates of the object. FT, Fourier transform. (Modified with permission from The Institute of Electrical and Electronics Engineers. Copyright 1984 IEEE [7])

proposed a method of measuring scattering at those angles for which $\vec{K}_0 \cdot \vec{K}$ is equal to zero and those for which it is equal to 1, thus providing enough equations to solve for both γ_κ and γ_ρ.

In general, density variations can be ignored in tissues because scattering due to density is less than that due to compressibility for objects of large extent [26].

3.6 Graphic Depiction of Scattering Classes

Scattering Classes 0 through 3 can be depicted in the Fourier domain as shown in Figure 3.9. The PSF is shown for a typical scanner along with the typical spectral ranges of various biologic components. One can see that the specular scatterers, that is, extended scatterers, are depicted only in a small region and only partially at best. In fact only the edge orthogonal to the beam can be detected because other edges have spectral characteristics that cannot be accessed by the PSF. The Class 1 (speckle) scatterers are not accessed because their spectral frequency characteristics are too high. They cause aliasing which produces speckle. The resolved scatterers, Class 2, are accessed in a small region because their spectrum is sampled with low frequency across the beam and high frequency axially along

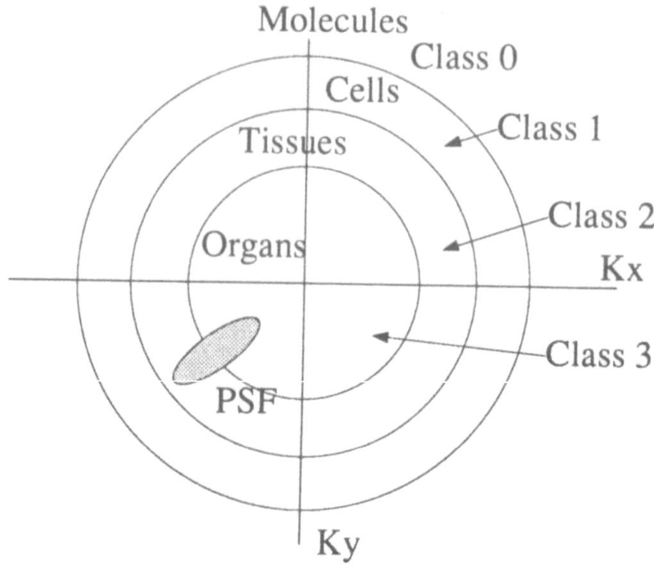

Figure 3.9. The point source function (*PSF*) of the imaging system defines the scattering classes available to the system. For typical medical ultrasound frequencies of 2.0 to 10.0 MHz, the scattering classes are arranged with the biologic classes as shown in this schematic of the Fourier domain.

the beam. The Class 0 scatterers produce a general scalar shift of the entire spectrum, which may be a weak function of spatial frequency. Probing the object from other directions results in the PSF moving around the center of the Fourier domain. This is much like compound scanning and results in better imaging because more of the frequency domain of the object is sampled. It is clear from Figure 3.9 that the biologic hierarchy of scattering classes depends on the position of the PSF in the Fourier domain. If cells are to be resolved, then higher frequencies and bandwidths are required. If the object is to be imaged completely, then the entire Fourier domain must be sampled by using many angles of view and many frequencies.

It is timely to mention here that the classes of scattering are associated with the PSF, thus the imaging system and not with the biologic class. It is only the fact that typical medical imaging instruments use frequencies ranging from 1 to 10 MHz that places the scattering classes in the same range as the biologic classes as stated in this book. If the PSF was stationed much farther out in the frequency range, then molecules and cells could be imaged as scattering Class 1 or Class 2. Indeed this is the case for ultrasonic microscopes[33].

3.7 Discussion

To the extent that Eqs. (3.6) and (3.7) represent the propagation of elastic waves in tissues, first-order Born and Rytov imaging can be accomplished in tissues. However, aberrations result if the second-order terms that have been dropped in the process of transforming Eqs. (3.6) and (3.7) into mathematically tractable equations are not negligible (underlined terms). One can say that the Born approximation results in depiction of propagation of the complex amplitude of the elastic pressure wave. It is a diffraction method because it is clear from Eq. (3.6) that the Born method predicts the same scattering distribution for identical objects that differ in scattering strength. The Rytov approximation is a depiction of the propagating complex phase of the elastic wave and also uses the Helmholtz equation but accounts for different scattering strengths of identically shaped objects. It is almost a consensus of the investigators in the field that transmission tomography experiments are best characterized with the Rytov approximation but that backscattering experiments are best characterized by the Born approximation.

The geometric relationships described in this chapter may be useful as a method of first approximation for calculating scattering from objects or for calculating objects from scattering. Second-order scattering might be approximated by using the highest-amplitude scattered waves as incident waves and repeating the procedure [34] in an iterative way.

It is the relentless increase in computer capabilities per unit cost that will enable the most complex inverse-scattering methods to be applied to biomedical imaging and that provides steady incentive for further study of the subject.

Bibliography

[1]   M. Kaveh, R. K. Mueller, and R. D. Iverson, "Ultrasonic tomography based on perturbation solutions of the wave equation," *Computer Graphics Image Process*, vol. 9, pp. 105–116, 1979.

[2]   S. A. Johnson, Y. Zhou, M. K. Tracy, M. J. Berggren, and F. Stenger, "Inverse scattering solutions by a sinc basis, multiple source, moment method–Part III: Fast algorithms," *Ultrasonic Imaging*, vol. 6, pp. 103–116, January, 1984.

[3]   P. L. Carson, T. V. Oughton, W. R. Hendee, and A. S. Ahuja, "Imaging soft tissue through bone with ultrasound transmission tomography by reconstruction," *Medical Physics*, vol. 4, pp. 301–309, 1977.

[4]   G. Glover and J. C. Sharp, "Reconstruction of ultrasound propagation speed: Distributions in soft tissue," *IEEE Transactions on Sonics and Ultrasonics*, vol. SU-24, pp. 229–234, 1977.

[5] S. A. Goss, R. L. Johnston, and F. Dunn, "Comprehensive compilation of empirical ultrasonic properties of mammalian tissues," *Journal of the Acoustical Society of America*, vol. 64, pp. 423–457, 1978.

[6] J. F. Greenleaf, "Three-dimensional imaging in ultrasound," *Journal of Medical Systems*, vol. 6, pp. 579–589, 1982.

[7] J. F. Greenleaf, "An inverse view of scattering," *Proceedings IEEE Ultrasonics Symposium*, vol. 2, no. 2, pp. 821–824, 1984.

[8] R. K. Mueller, "Diffraction tomography. I: The wave-equation," *Ultrasonic Imaging*, vol. 2, pp. 213–222, July, 1980.

[9] G. T. Herman, *Image Reconstruction from Projections: The Fundamentals of Computerized Tomography*. Academic Press. New York, 1980.

[10] E. Wolf, "Three-dimensional structure determination of semi-transparent objects from holographic data," *Optical Communication*, vol. 1, pp. 153–156, September/October, 1969.

[11] K. Iwata and R. Nagata, "Calculation of refractive index distribution from interferograms using Born and Rytov's approximation.," *Japanese Journal of Applied Physiology*, vol. 14(Suppl. 14, no. 1), pp. 379–383, 1975.

[12] A. J. Devaney, "A filtered backpropagation algorithm for diffraction tomography," *Ultrasonic Imaging*, vol. 4, pp. 336–350, October, 1982.

[13] J. F. Greenleaf, "Computed tomography from ultrasound scattered by biological tissue," in *Inverse Problems* (D. W. McLaughlin, ed.), vol. 14, (American Mathematical Society. Providence, RI), pp. 53–63, 1984.

[14] R. K. Mueller, M. Kaveh, and G. Wade, "Reconstructive tomography and applications to ultrasonics," *Proceedings of the IEEE*, vol. 67, pp. 567–587, 1979.

[15] D. Nahamoo, S. X. Pan, and A. C. Kak, "Synthetic aperture diffraction tomography and its interpolation-free computer implementation," *IEEE Transactions on Sonics and Ultrasonics*, vol. SU-31, no. 4, pp. 218–229, July, 1984.

[16] R. Dandliker and K. Weiss, "Reconstruction of three-dimensional refractive index from scattered waves," *Optical Communication*, vol. 1, no. 7, pp. 323–328, 1970.

[17] S. A. Johnson, F. Stenger, C. Wilcox, J. Ball, and M. J. Berggren, "Wave equations and inverse solutions for soft tissue," in *Acoustical Imaging* (M. Kaveh, R. K. Mueller, and J. F. Greenleaf, eds.), vol. 11, pp. 409–424, Plenum Publishing Corporation, New York, 1982.

[18] M. Kaveh, M. Soumekh, and R. K. Mueller, "A comparison of Born and Rytov approximations in acoustic tomography," in *Acoustical Imaging* (M. Kaveh, R. K. Mueller, and J. F. Greenleaf, eds.), vol. 11, pp. 325–335, Plenum Press, New York, 1982.

[19] S. Wanuga (Editor-in-Chief), "Special issue on digital acoustical imaging," *IEEE Transactions in Sonics and Ultrasonics*, vol. SU-31, no. 4, pp. 193–418, 1984.

[20] *Wave Motion 11*. Elsevier Science Publishers B.V. North-Holland, Amsterdam, 1989.

[21] P. M. Morse and H. Feshbach, *Methods of Theoretical Physics*. McGraw-Hill. New York, 1953.

[22] E. L. Carstensen, W. K. Law, N. D. McKay, and T. G. Muir, "Demonstration of nonlinear acoustical effects at biomedical frequencies and intensities," *Ultrasound in Medicine and Biology*, vol. 6, pp. 359–368, 1980.

[23] F. Dunn, W. K. Law, and L. A. Frizzell, "Nonlinear ultrasonic propagation in biological media," *British Journal of Cancer*, vol. 45(Suppl. 5), pp. 55–58, 1982.

[24] C. M. Sehgal and J. F. Greenleaf, "Scattering of ultrasound by tissues," *Ultrasonic Imaging*, vol. 6, pp. 60–80, January, 1984.

[25] G. H. Brandenburger, J. R. Klepper, J. B. Miller, and D. L. Synder, "Effects of anisotropy in the ultrasonic attenuation of tissue on computed tomography," *Ultrasonic Imaging*, vol. 3, pp. 113–143, April, 1981.

[26] B. S. Robinson and J. F. Greenleaf, "Measurement and simulation of the scattering of ultrasound by penetrable cylinders," in *Acoustical Imaging* (M. Kaveh, R. K. Mueller, and J. F. Greenleaf, eds.), vol. 13, pp. 163–178, Plenum Publishing Corporation, New York, 1984.

[27] A. J. Devaney, "Inverse source and scattering problems in ultrasonics," *IEEE Transactions on Sonics and Ultrasonics*, vol. SU-30, no. 6, pp. 355–364, 1983.

[28] M. Kaveh, M. Soumekh, and J. F. Greenleaf, "Signal processing for diffraction tomography," *IEEE Transactions on Sonics and Ultrasonics*, vol. SU-31, no. 4, pp. 230–238, July, 1984.

[29] J. B. Keller, "Signal processing for diffraction tomography (Letter to the editor)," *Journal of Acoustical Society of America*, vol. SU-31, pp. 1003–1004, 1969.

[30] R. F. Wagner, S. W. Smith, J. M. Sandrik, and H. Lopez, "Statistics of speckle in ultrasound B-scans," *IEEE Transactions on Sonics and Ultrasonics*, vol. 30, pp. 156–173, 1983.

[31] J. F. Greenleaf and A. Chu, "Multifrequency diffraction tomography," in *Acoustical Imaging* (M. Kaveh, R. K. Mueller, and J. F. Greenleaf, eds.), vol. 13, pp. 43–55, Plenum Publishing Corporation New York, 1984.

[32] S. J. Norton and M. Linzer, "Ultrasonic reflectivity imaging in three dimensions: exact inverse scattering solutions for plane, cylindrical, and spherical apertures," *IEEE Transactions in Biomedical Engineering*, vol. BME-28, pp. 202–220, 1981.

[33] E. A. Ash, *Scanned Image Microscopy*. Academic Press. London, 1980.

[34] M. Azimi and A. C. Kak, "Multiple scattering and attenuation phenomena in diffraction imaging," tech. rep., School of Electrical Engineering, Purdue University, West Lafayette, Indiana (TR-EE 85-4), February, 1985.

4
Class 0 Scattering

4.1 Introduction

It was pointed out in Chapter 1 that ultrasound interacts intimately with tissues. These interactions occur at different levels, which from a practical point of view can be classified on the basis of the sizes of the acoustic inhomogeneities. In principle, the interactions occur at molecular as well as macroscopic levels.

4.1.1 Molecular Interactions

Molecular or biomolecular interactions are usually associated with sound propagation properties like absorption and acoustic speed, the Class 0 scatterer. In general, for absorption of ultrasound to occur, the system should possess forms of equilibrium that can be perturbed by pressure. On the other hand, the finite time taken by the sound waves to travel from one point to another requires "spring-like" compressible bonds joining the rigid molecules. Based on the pressure amplitude of sound waves, these bonds could behave as either harmonic or anharmonic oscillators and cause the medium to respond linearly or nonlinearly, respectively, to the compressions and the rarefactions of sound waves.

4.1.2 Macroscopic Interactions

The macroscopic subset of the tissue hierarchy consists of scatterers of Classes 1, 2, and 3. The scattering behavior of such systems depends on the size of the particles compared to the wavelength of the interrogating wave. The inhomogeneities that are significantly smaller than the wavelength lead to Rayleigh scattering; the inhomogeneities comparable to the wavelength cause ultrasound to diffract; and the inhomogeneities significantly greater than the wavelength behave like polished surfaces and the propagation occurs according to geometric principles.

4.1.3 Tissues

It is important to note that in tissues all the interactions listed above are observed in addition to interactions with moving scatterers or Class 4 scatterers, described in Chapter 6. This means that the acoustic signal after propagation through a medium contains a wealth of information which potentially can be used to identify and monitor disease processes. However, the simultaneous occurrence of these phenomena makes the decoding of the information difficult and has been the subject of intense investigation. Although there are many questions that remain to be answered, considerable progress has been made over the last decade toward the understanding of biophysics of sound propagation through tissues and the exploitation of these principles for beneficial purposes.

The discussion that follows investigates the different ways in which ultrasound interacts with tissues. Emphasis is on the various models that are used to interpret and evaluate intrinsic properties of tissues from the acoustic measurements. This chapter deals with the phenomena associated with Class 0 scatterers: absorption of ultrasound by tissues, speed of sound propagation, and acoustic nonlinearity. Chapter 5 deals with interactions of sound waves with the scatterers of Classes 1, 2, and 3 and Chapter 6 deals with motion detection.

4.2 Class 0 Scatterers

4.2.1 Absorption of Ultrasound

The phenomenon of absorption is defined as the transformation of ultrasonic energy to heat. To date there is no fully developed model that can comprehensively explain all the experimental observations for the absorption of ultrasound by tissues. Some models treat propagation of sound waves through tissues as similar to that of liquids, others compare the behavior of tissue to that of semisolids, and still others treat tissue propagation as suspensions of inhomogeneous particles. Although each of the models emphasizes different interactions, a common thread, i.e., relaxation phenomenon, is observed in all the cases in some form or other.

The literature on absorption is vast and all aspects are not covered here. The scope of the discussion is limited to normal physiologic conditions and to the conditions that are currently amenable to imaging and tissue characterization processes. With this in mind the discussion is limited to solid soft tissues irradiated with longitudinal waves of low intensity in the frequency range of 1 to 10 MHz.

Definitions. The attenuation of a signal is thought to arise largely from two mechanisms: absorption and spatial redistribution of acoustic energy. The latter is often referred to as the "scattering" component. From a historical perspective, there has been some controversy as to how much each component contributes to the total. Currently, the consensus leans toward absorption being the dominant mechanism (>90%–95%), at least in tissues like liver and brain [1].

There are three types of definitions encountered in the literature for absorption coefficient, α.

$$\alpha_{excess} = \frac{\alpha_{solution} - \alpha_{solvent}}{[solute]} \qquad (4.1)$$

$$\alpha_{specific} = \frac{\alpha_{measured}}{\rho_{solution}} \qquad (4.2)$$

$$\alpha_{nonclassical} = \alpha_{measured} - \alpha_{classical} \tag{4.3}$$

Excess and specific absorptions enable comparison of the absorption properties independent of concentration of solute and density of solution, respectively. Nonclassical absorption, on the other hand, refers to the absorption not accounted for by classical mechanisms of energy loss. Often absorption is expressed as frequency (f)-independent quantities, α/f or α/f^2, or as $\alpha\lambda$, which is absorption per unit wavelength (λ).

Mechanisms of Absorption   Transformation of acoustic energy to heat by frictional forces and heat conduction is called classical absorption and is given by the equation

$$\alpha_{classical} = \frac{\omega^2}{2\rho c^3}\left[\frac{4}{3}\eta_{shear} + \eta_{bulk} + \frac{\gamma - 1}{c_p}K\right], \tag{4.4}$$

where η_{shear} and η_{bulk} represent shear and bulk viscosities, respectively; γ is the specific heat ratio; ρ is the density; c_p is the specific heat of the medium at constant pressure; K is the thermal conductivity; and ω is the angular frequency.

Even though fluids and soft tissues do not support static shear, they can support a dynamic one in the form of viscous drag. This gives rise to a shear term in the Eq. (4.4). The last term in the brackets represents the contribution of energy loss by heat conduction from the regions of high temperatures (compressed zones) to the neighboring regions of lower temperatures (rarified zones). For this mechanism to play a significant role, thermal conductivity, K, should be large. If one assumes the value of K for soft tissues to be close to that of water, the last term for tissues is small and can be neglected.

On the basis of Eq. (4.4) one can make two predictions. First, absorption should increase quadratically with insonation frequency, ω; and second, the absorption should show temperature dependence similar to that of viscosity, η, terms. To see how these expectations stand up to experiment, consider the data shown in Figure 4.1, which summarizes the results from various laboratories.

Based on the experimental results several generalizations can be made. Attenuation (or absorption assuming scattering "loss" to be negligible) in biologic media is about 100 times that of water, even though water is the major component. Absorption increases almost linearly with frequency in the range of 1 to 10 MHz as opposed to the quadratic dependence exhibited by water. Speed dispersion in tissues like brain is very small, on the order of -1 m/sec/MHz. Alpha shows a complex temperature dependence which varies with insonation frequency. Also, absorption changes with molecular size; the critical transition occurs around a molecular weight of 500.

To explain some of the observations and, in particular, their relatively high value of absorption and linear frequency dependence (the two facts that directly impact current imaging and tissue characterization techniques) several models have been proposed which view tissues as liquids, semisolids, or suspensions. The consequence of each consideration is discussed below.

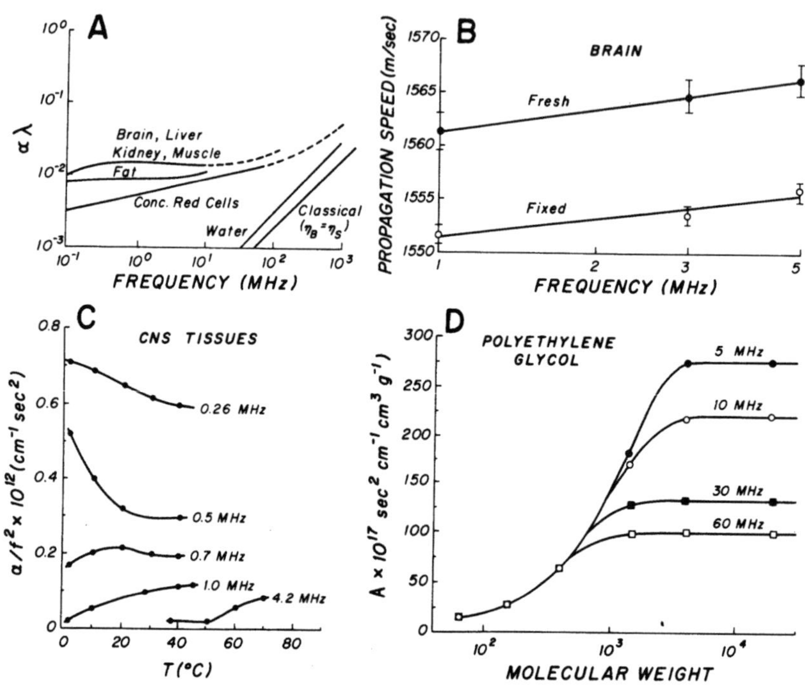

Figure 4.1.   (A) Absorption per wavelength of ultrasound in soft tissues. The dashed lines are the extrapolated estimates.   (From [2]) (B) Speed of sound propagation in fresh and fixed tissues from normal adult brain.   (Reprinted with permission from Acoustical Society of America [3]) (C) Temperature dependence of ultrasonic absorption in mammalian central nervous tissue at different frequencies of insonation.   (Reprinted with permission from Acoustical Society of America [4]) (D) Change in absorption with molecular weight of random-coil polymer, polyethylene glycol. Open and filled circles and squares represent the measurements made at 5, 10, 30, and 60 MHz ultrasound. (Reprinted with permission from L. W. Kessler, W. D. O'Brien, Jr., and F. Dunn, "Ultrasonic absorption in aqueous solutions of polytheylene glycol," J. Phys. Chem., vol. 74, pp. 4096–4102, 1970. Copyright 1970 American Chemical Society [5])

Tissues as Liquidlike.   According to the model that treats tissues as liquids or solutions, the total absorption is due to classical mechanisms and, to a larger degree, due to the various chemical relaxation processes. These processes arise as a result of chemical equilibria like proton transfer, solute-solvent interactions, conformation changes, and macromolecular interactions.

To determine the degree to which absorption occurs at macroscopic or molecular levels, Pauly and Schwan [6] measured attenuation, using a substitution method, of various preparations of homogenized beef liver and saline. The experimental results and the conclusions that can be drawn from them are described as follows.

1. Destruction of gross structure of a tissue by homogenization results in only moderate (33%) reduction in absorption. Recently, researchers have found

this component to be even smaller [1]. This leads one to infer that a major fraction of absorption occurs at a "submicroscopic," probably macromolecular, level.

2. The frequency dependence of absorption is characterized by a power function f^n, where $n = 1$ to 1.4. Such behavior cannot be explained on the basis of classical models defined by Eq. (4.4). Without specifying the nature of equilibria involved, Pauly and Schwan [6] explained this behavior as due to multiple molecular relaxations that occur during sound propagation. Such relaxations were described by the equation

$$\frac{\alpha\lambda}{\omega} = A + 2 \int_{T_1}^{T_2} \frac{(\alpha\lambda)^* T}{1 + \omega^2 T^2} dT, \tag{4.5}$$

where A represents classical absorption and the integral represents a continuum of relaxation processes, with relaxation times ranging from T_1 to T_2. If $(\alpha\lambda)^*$ is a function of T defined by the equation

$$(\alpha\lambda)^* = a^{\#} T^n, \tag{4.6}$$

where a^* is the constant of proportionality and n is an integer, combining Eq. (4.5) and Eq. (4.6),

$$\alpha\lambda = \omega A + \frac{2a}{\omega^{n+1}} \int_{\omega T_1}^{\omega T_2} \frac{(\omega T)^{n+1}}{1 + (\omega T)^2} d(\omega T). \tag{4.7}$$

Pauly and Schwann [6] calculated absorption per cycle by the above equation for various values of n. The absorption per cycle versus f curves for different values of n and different ranges of T_1 and T_2 are summarized in Figure 4.2. In particular, for the case $n = -1$ and $T_2/T_1 = 1000$, such a model predicts high absorption and linear frequency dependence from 0.1 to 10 MHz.

3. Absorption of ultrasound by tissues changes significantly with the denaturation of proteins, thereby leading to the inference that the macromolecular absorption is associated mainly with proteins. If this is the case, the question that follows is, what types of proteins are they?

Most of the data on the absorption as a function of pH, as shown in Figure 4.3, illustrate that proteins have high absorption at acidic and basic pH values and relatively weak absorption around neutral pH values. This means that if proteins are significant contributors to ultrasonic absorption, they should contribute in significant measure at values close to neutral pH.

Slutsky et al. [7] demonstrated that the proteins with histidyl residue can cause absorption around neutral pH values via proton transfer reaction of the type

$$-I + H_2PO_4^- \underset{k_b}{\overset{k_f}{\rightleftharpoons}} -IH^+ + HPO_4^{2-}, \tag{4.8}$$

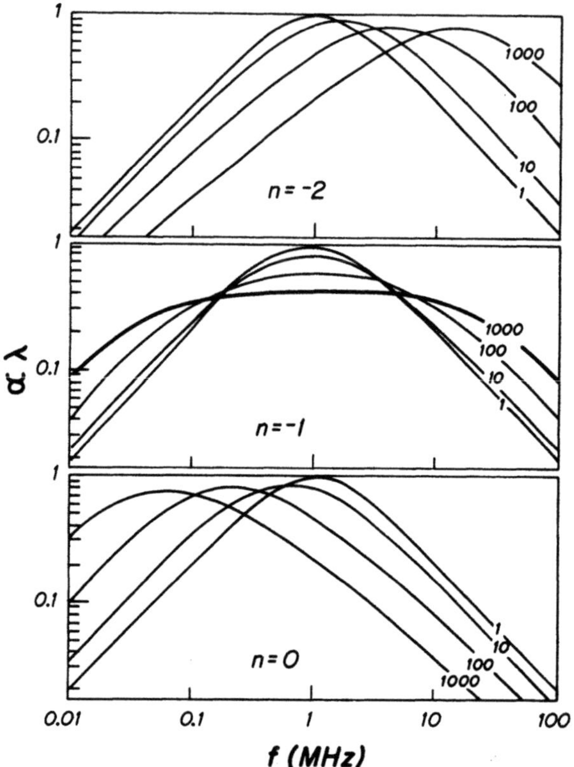

Figure 4.2. Absorption per wavelength based on a multiple relaxation model. Equation (4.7) was used for the calculation of absorption in excess of contributions due to the term A. The three panels represent different time-constant functions defined by Eq. (4.6) for the values of -2, -1, and 0 for n. Families of curves in each panel correspond to time-constant ratios (T_2/T_1) from 1 to 1000. (Reprinted with permission from Acoustical Society of America [6])

where $-I$ represents a histidyl residue or an N-terminal amino acid of a protein.

For such an equilibrium reaction, the nonclassical absorption is given by [7],

$$\frac{\alpha}{f^2} - A = \frac{(C/f_r)}{1 + (f/f_r)^2},\tag{4.9}$$

where

$$C \simeq \frac{2\pi^2 \rho c \Gamma (\Delta V)^2}{RT}.\tag{4.10}$$

In Eq. (4.9) and (4.10), ΔV is change in volume associated with equilibrium, R is the gas constant, and T is absolute temperature. The constant Γ is determined by the stoichiometry of the reaction equation (Eq. [4.8]). The

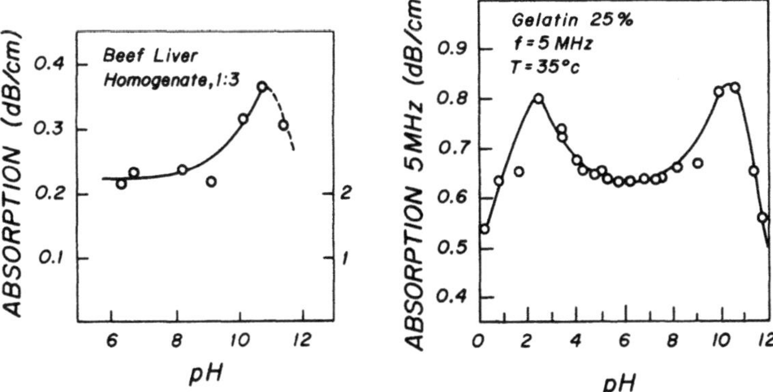

Figure 4.3. Dependence of absorption coefficient on the pH of the medium. Left panel: Absorption at 1 MHz in 33% beef liver homogenate (25°C). Right panel: Absorption at 5 MHz in 25% gelatin solution (35°C). (Reprinted with permission from Acoustical Society of America [6])

resonance frequency f_r is determined from the rate constant of the forward reaction and the concentrations of inorganic phosphates and the titratable histidyl residues. A substantial volume change ($\Delta V = -25$ ml) is associated with the chemical equilibrium represented by Eq. (4.8). This makes C significant in magnitude and thus leads to a high absorption coefficient.

If one considers tissues to be composed of proteins with N sites for proton exchange, the overall rate is described by a family of chemical equilibria similar to the type described by Eq. (4.8). By solving the linearized rate equations of the various equilibria the relaxation frequency of the proton exchange between proteins and buffer is derived [7],

$$2\pi f_r = k_f(Nc_0 + c_b), \tag{4.11}$$

where Nc_0 represents the concentration of sites of proton exchange and c_b is the buffer concentration.

With a series of assumptions made about the amount of intra- and extracellular phosphates present in tissues, and the fraction that is titratable at neutral pH, the magnitude of Nc_0 and c_b is determined to compute C and f_r by Eqs. (4.10) and (4.11). The two values, when used in Eq. (4.9), yield absorption coefficients. The estimated values are summarized in Figure 4.4.

The hatched area represents the estimated values of absorption for soft tissues. The lower range was determined by assuming that the histidyl residue titrates 1 pH unit on the acid side of neutrality. These results illustrate that the high absorption coefficient of tissues can be predicted on the basis of a proton transfer reaction involving histidyl residues. However, it does not fully

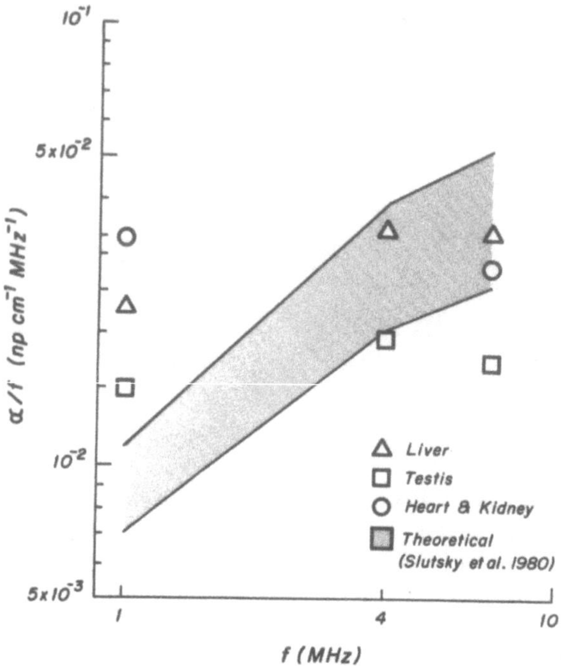

Figure 4.4.   Comparison of the experimental values [8] for the absorption coefficient with the values we calculated from the model proposed by Slutsky et al. [7]. Circles, triangles, and squares represent the experimental data [8].

predict linear frequency dependence. Nevertheless, it can be inferred that the differences in the absorption value of various tissues are due to differences in mole-fraction of histidine present in the medium. To test the hypothesis that "the differences in protein absorption are due to the differences in the histidine content," Kremkau [9] extensively studied the absorption of ultrasound by various proteins in phosphate buffers. Table 4.1 summarizes the results on the fraction of total absorption attributable to the histidine component of the proteins.

Table 4.1 Comparison Between the Absorption Coefficients Predicted on the Basis of Histidine Concentration and the Total Measured Absorption (at 30 MHz) by Proteins in Phosphate Buffer.[1]

Protein	% Histidine	Calculated absorption due to histidine, cm^2/g	Measured total specific absorption coefficient, cm^2/g	% Protein absorption accounted for by histidine residue
Albumin	3.00	1.38	9.49	15
Bacitracin	9.09	4.17	8.87	47
γ-Globulin	1.68	0.77	6.34	12
Hemoglobin	6.62	3.04	7.97	38
Hexokinase	2.13	0.98	6.36	15
β-Lacto-globulin	1.23	0.56	7.34	8
Lysozyme	0.78	0.36	4.66	8
Urease	2.65	1.22	5.40	23

[1]Obtained by multiplying specific absorption coefficient of histidine in phosphate buffer (45.9 cm^2/g at 30 MHz) and histidine concentration in the protein listed in the second column [9].

It is clear from these results that even in the relatively simple systems, histidyl residue can account for not more than 10%–50% of the total absorption. Also, there is a poor correlation between the measured absorption and that predicted on the basis of protein concentration. This, at least in part, could be due to simplifying assumptions used in calculating histidine contributions, namely, that proteins are of equal molecular weights and that the histidine absorption is the same in proteins as it is in solutions. Nevertheless, the result of this study effectively demonstrates that additional molecular mechanisms which take into consideration the solute-solvent interactions are also operational in significant measure in tissues.

In short, it is reasonable to conclude that the models based on chemical relaxation processes (multiple or single) have some attractive features. They enable us to explain some salient experimental features but as yet have not reached quantitative precision to find wide application for determining composition of tissues and thus enabling their characterization.

Tissues as Semisolids.   An alternative approach treats tissues as semisolids. That is to say, the stress they experience is not only related to strain but also to the

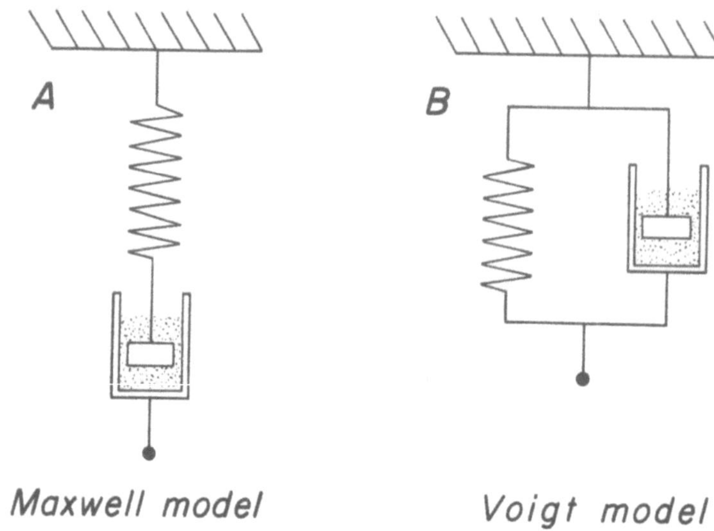

Figure 4.5. (A) The Maxwell element. (B) The Voigt model.

rate at which strain is applied. Such a medium, often referred to as viscoelastic, is equivalent to combinations of dashpots and springs arranged either in series (Maxwell model) or in parallel (Voigt model). These arrangements are shown in Figure 4.5.

The damping and elastic properties of dashpot and spring, respectively, are expressed by the following stress-strain relationships.

1. Dashpot

$$Stress = viscosity * strain\ rate, \tag{4.12}$$

or

$$\sigma_{dashpot} = \eta\left(\frac{\partial \epsilon}{\partial t}\right) \tag{4.13}$$

2. Spring

$$Stress = elasticity\ constant * strain \tag{4.14}$$

or

$$\sigma_{spring} = E\epsilon \tag{4.15}$$

For soft tissues, the Voigt model is assumed. That is, the strain is constant and stresses add. The net stress σ is given by the sum of Eqs. (4.13) and (4.15),

$$\sigma = \eta(\frac{\partial \epsilon}{\partial t}) + E\epsilon. \tag{4.16}$$

Eq. (4.16) is solved for sinusoidally varying stress and for shear deformations (i.e., substituting E by shear modulus, G) to yield frequency-dependent complex moduli [10].

$$G = G' + iG'\omega T_s \tag{4.17}$$

where G' and T_s represent storage moduli and shear relaxation time, respectively. Note that loss moduli (imaginary component) is a function of G'. Similar approaches, with few alterations, can be used for compressional deformation to obtain a frequency-dependent bulk modulus, K [10].

$$K = K' + i\frac{K_2\omega T_b}{1 + (\omega T_b)^2}, \tag{4.18}$$

where K' is static bulk modulus as $\omega \to 0$, K_2 is the relaxation modulus, and T_b is the bulk relaxation time.

The moduli K and G are related to the sound speed by the equation

$$c_{RI}^2 = K + (4/3)G \tag{4.19}$$

where subscript RI indicates c is a complex number. Substituting the expressions for K and G in the last equation and comparing the real and the imaginary parts of the resulting equation with those of the following equation,

$$\frac{1}{c_{RI}} = \frac{1}{c} + i\frac{\alpha}{\omega}, \tag{4.20}$$

one obtains

$$c = \frac{1}{\rho}\left[K' + \frac{4}{3}G'\right] \tag{4.21}$$

and

$$\frac{\alpha}{f} = \frac{\pi}{\rho c^3}\left[\frac{K_2\frac{f}{f_v}}{1 + (\frac{f}{f_v})^2} + \frac{4}{3}G'\frac{f}{f_v}\right]. \tag{4.22}$$

In Eqs. (4.21) and (4.22) the quantity f_v is equal to $1/(2\pi T_v)$. In deriving these equations it is assumed that $T_b = T_s = T_v$ and that $\omega^2 >> \alpha^2 c^2$.

Ahuja [11] fitted absorption versus frequency data to Eq. (4.22) and determined the values of f_v, K_2, and G' to be 2 MHz, 2.2×10^8 ergs/cm^2, and 0.13×10^8 ergs/cm^2, respectively. With these values the two loss moduli

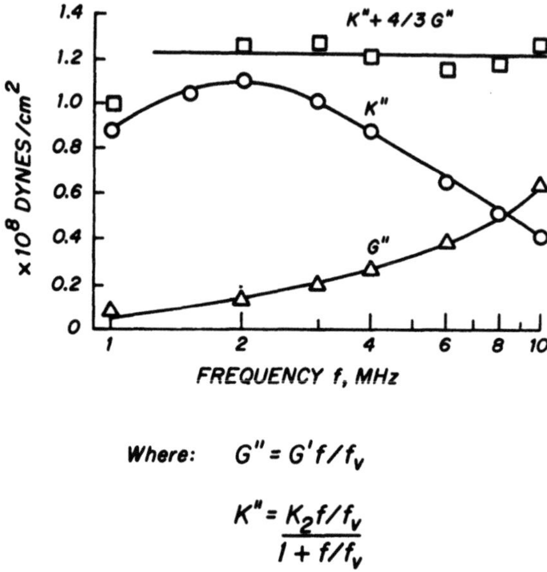

Figure 4.6. Prediction of absorption coefficient per cycle on the basis of model described. (Reprinted with permission from Academic Press [11])

K'' and G'', represented by the two terms in brackets in Eq. (4.22), were calculated. Plots of K'' and G'' and their sums are shown in Figure 4.6.

The variable K'' goes through a maximum at 2 MHz, whereas G'' increases monotonically with frequency. The sum of the two remains almost linearly constant, implying linear frequency dependence for tissues. However, the plot of α/f increases rapidly with frequency above 10 MHz (see Fig. 4.7). The experimental data suggest this is not the case [12]. One of the problems with this approach is that the experimental data were used to determine the values of K_2, G', and f_v, and then these numbers were used to explain a similar type of data. Independent measurement of these constants may, in the future, provide validity to this model. However, the magnitude of f_v (i.e., 2 MHz) obtained from viscoelastic models compares closely to the resonance frequencies calculated by Slutsky and co-workers [7]: 1.2 to 1.35 MHz for bacitracin and 1.6 to 2.9 MHz for muscle tissues. This raises the question if there is any relationship between the two apparently different approaches. To what extent this is true remains to be tested. However, in the classical limit: $f/f_v \ll 1$ and the denominator of the first term of Eq. (4.22) reduces to unity. Because $K_2/f_v = 2\pi\eta_{bulk}$ and $G'/f_v = 2\pi\eta_{shear}$, Eq. (4.22) reduces to

$$\frac{\alpha}{f^2} = \frac{2\pi^2}{\rho c^3}\left[\eta_{bulk} + \frac{4}{3}\eta_{shear}\right], \tag{4.23}$$

which is identical to classical absorption.

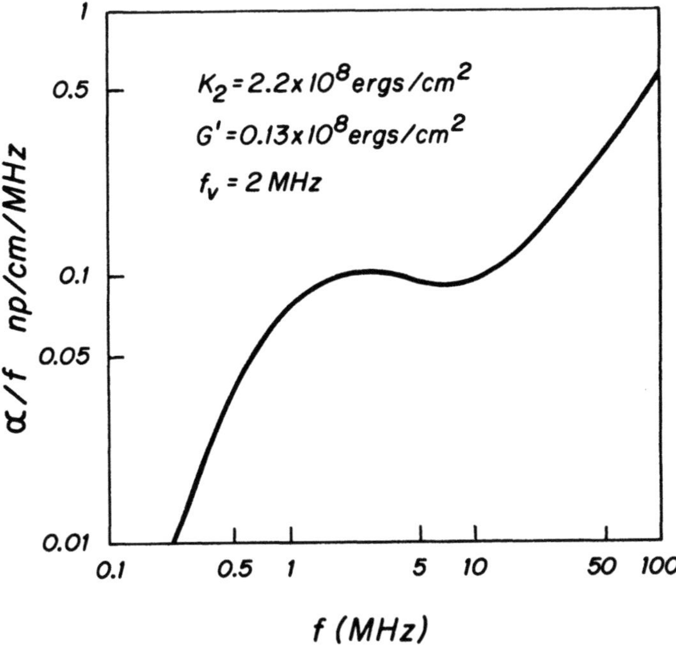

Figure 4.7.   Frequency behavior calculated by us from viscoelastic model [11].

Tissues as Suspensions.   Another mechanism that can potentially cause absorption in tissues can be best illustrated by treating a medium as a "suspension" in which relative motion between the particles and the embedding medium causes viscous damping of sound waves. In addition to this there is heat conduction between the suspended particles and the surrounding medium. The damping due to the thermal mechanism is believed to be small compared to that due to viscosity [13]. An analysis that uses the concepts of frictional force constants and equivalent masses yields a "relaxation-type" expression for absorption coefficient α_v [14],

$$\frac{\alpha_v \lambda}{V_p} = \pi \frac{m_e}{m_p} \frac{(1 - \rho_0/\rho_e)^2}{(\rho_0/\rho_e)} \frac{(\omega/\omega_0)}{1 + (\omega/\omega_0)^2}. \qquad (4.24)$$

In the above equations ρ_0 and ρ_e represent the density of the embedding medium and the structured elements, respectively. The symbols m_e and m_p are the mass and effective mass of the structural elements. V_p is the volume concentration of the suspended particles. If m is the mass of fluid displaced by a particle and we assume it to be spherical of radius a, m_p, and ω_0 can be defined by the equation

$$m_p = m_e + m \left[\frac{1}{2} + \frac{9}{4\delta} \right], \qquad (4.25)$$

Table 4.2 Percent of Scattering in Attenuation for Three Tissues.

Tissue	Scatter size, micrometer	Viscous absorption due to relative motion, cm/MHz	Experimental attenuation, cm/MHz^{-1}	Percent accounted
Blood	4.6 - 5.6	0.0003	0.031	10
Heart (muscle, myofibril)	1.0 - 2.0	0 .042	0.072	60
Skin (collagen fibrils)	0.7 - 1.5	0.101	0.170	60

and

$$\omega_0 m_p = 6\pi a \eta_{shear}[1 + \delta],\tag{4.26}$$

where

$$\delta = \left(\frac{\rho_0}{2\eta_{shear}}\right)^{\frac{1}{2}} a\omega^{\frac{1}{2}}.\tag{4.27}$$

The form of relaxation curve expressed by Eq. (4.24) is broader than that of the previously mentioned relaxation processes: therefore, perhaps it will give a broader and flatter relationship of absorption per cycle versus frequency [15]. O'Donnell and Miller [13] calculated the losses due to this mechanism, given the scatter sizes and their relative densities. This study estimates viscous absorption due to relative motion to be about 10% to 60% of the experimentally measured attenuation (Table 4.2). This mechanism predicts some additional losses in tissues that are the result of inhomogeneity of tissues.

The significance of this mechanism to the total has been reevaluated recently in a series of studies conducted by Miles and co-workers [16, 17]. They modeled myofibrils as prolate spheres and derived a modified equation for viscous absorption,

$$\frac{\alpha_v \lambda}{V_p} = \frac{\pi\left(\frac{\rho_e}{\rho_0} - 1\right)S}{2\left[\left(\frac{\rho_e}{\rho_0} + \tau'\right)^2 + S^2\right]},\tag{4.28}$$

where τ' and S are referred to as shape factor and inertia coefficient, respectively. The calculations based on this formula show that absorption by the viscous mechanism does not account for more than 10% to 15% of the total, which is significantly lower than the earlier estimates. These calculations were further confirmed experimentally by measuring components of attenuation that arise due to viscous mechanisms.

Other Models.   From the studies described above it is rather clear that each of the models has emphasized relaxation processes in some form or other. In the majority of cases the relaxation processes have not been observed directly. This is probably because the various relaxation processes are not easily separable from one another. It is quite likely that most of the mechanisms participate and may, to some degree, be interrelated on a molecular level.

A model that did not look at the individual relaxation processes but instead at how such processes affect the thermodynamics of propagation was proposed by Sehgal and Greenleaf [18]. There were two implicit assumptions made.

1.   Various processes (thermal, viscous, or chemical) that contribute to absorption can be represented by a general equation,

$$\frac{d\mu}{dt} = \frac{1}{T_R}(\mu - \mu_0),\tag{4.29}$$

where T_R is the relaxation time and μ represents the individual process.

2.   Each of these relaxation processes makes the sound propagation less adiabatic and irreversible and leads to a time (or phase) delay between the pressure wave and the accompanying thermal wave.

To account for the loss of adiabatic character, the exponent of specific heat ratio, γ, was expressed in the wave equation by a complex number, where the imaginary part represents the loss term.

$$\rho\frac{d^2p}{dt^2} = \frac{\gamma}{\beta_T}\frac{d^2p}{dx^2}\tag{4.30}$$

or

$$\rho\frac{d^2p}{dt^2} = \frac{\gamma^{1+ib}}{\beta_T}\frac{d^2p}{dx^2}\tag{4.31}$$

Solving the above equation and assuming that the time delays due to each mechanism sum, it was shown that,

$$\alpha = \frac{\omega}{c}sin\left[\frac{\omega\tau ln(\gamma)}{2(1+\omega^2\tau^2)^{1/2}}\right],\tag{4.32}$$

where τ is the total time delay.

In the last equation there are two unknowns, τ and γ. To see how well this equation tracks the frequency dependence, two examples are shown in Figure

                    J. F. Greenleaf and C. M. Sehgal

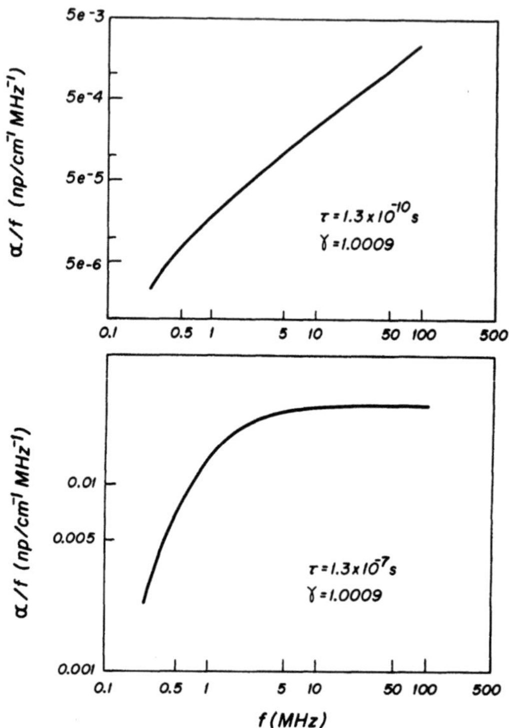

Figure 4.8.   Prediction of absorption coefficient on the basis of model from [18].

4.8 with different values of τ and assuming a value of γ close to unity, a value expected for liquids and tissuelike media.

When τ is small, α/f increases with frequency as is observed for fluids (top panel of Figure 4.8). On the other hand, for large values of τ, α/f stays constant for a wide frequency range, thus predicting linear frequency dependence (bottom panel). Intermediate values of τ would yield frequency dependence between 1 and 2, as is observed in tissues. Although this model predicts frequency dependence and also a large magnitude of absorption, it is at the expense of not having any information on the individual mechanism that may contribute to the total absorption.

There are a few other models that have not been considered here: one by Leeman [19] in which a new wave equation with a loss term was proposed. The model has yet to be tested as to its prediction of scattering and wave propagation in tissue. Arthur and Gurumurthy [20] modeled tissue as a transfer function with single pole zero combination and have found good fits of their model to data. The advantage of this model is that it also predicts phase velocity dispersion reasonably well. Jongen et al. [21] proposed a model in which several relaxation models are combined into one, which predicts the distribution of absorption fairly well.

Conclusions.   So far we have reviewed various models. Each of these models explains specific aspects of absorption in tissues. As yet there is no unified model that provides a comprehensive picture of the phenomenon. This fact, along with other technical complications, has made the use of this variable for the purposes of tissue characterization a challenging job. Although a wide range of mechanisms is consistent with some of the measured data, problems exist: 1) all the data are not predicted by the current models; 2) there is a need for a series of experiments to specifically separate the various contributing mechanisms; and 3) adequate measurements of absorption in tissues over a wide enough range of frequencies need to be made.

4.2.2 Sound Speed and Acoustic Nonlinearity (B/A)

Another set of variables that is used for nondestructive biologic system evaluation includes the sound speed and the acoustic nonlinearity variables, B/A. Measurement of these variables provides a means for obtaining equilibrium thermodynamic properties that are not readily determined by other imaging modalities used in medicine. For example, adiabatic compressibility, β_S, which is related to compliance of tissues, can be determined by using the measured sound speed, c, and the density, ρ, in the following hydrodynamic formula,

$$\beta_S = \frac{1}{c^2 \rho}. \tag{4.33}$$

One of the factors that determines the magnitude of compressibility of a medium is the space between the constituent molecules. This space in turn is a function of intermolecular forces. Because sound speed is related to compressibility, it is reasonable to infer that the former can provide information regarding the intermolecular bonding. As many physiologic and possibly pathologic problems concerning tissues may be related intermolecular biochemical processes, the study of this parameter to characterize tissues is of interest.

On a molecular scale one can regard tissues as a complex matrix of biomolecules which are represented by large circles enclosing smaller "molecular" circles in Figure 4.9. Such molecular systems are interlaced by intermolecular bonds like hydrogen bonds or bonds due to weak Van der Waals forces (represented by thick lines). Sound propagation through such media occurs at infinite speed through molecules (rigid incompressible spheres) and with finite time delay from one molecule to another through compressible intermolecular bonds. Depending on the amplitude of the acoustic waves these bonds could oscillate harmonically or anharmonically, making a medium respond linearly or nonlinearly, respectively, to pressure perturbations. The two limiting cases are shown in Figure 4.10.

If tissues behave like three-dimensional harmonic oscillators, equal pressures lead to equal compressions or rarefactions. That is, pressure versus volume is a linear function, the slope of which is constant and related to compressibility.

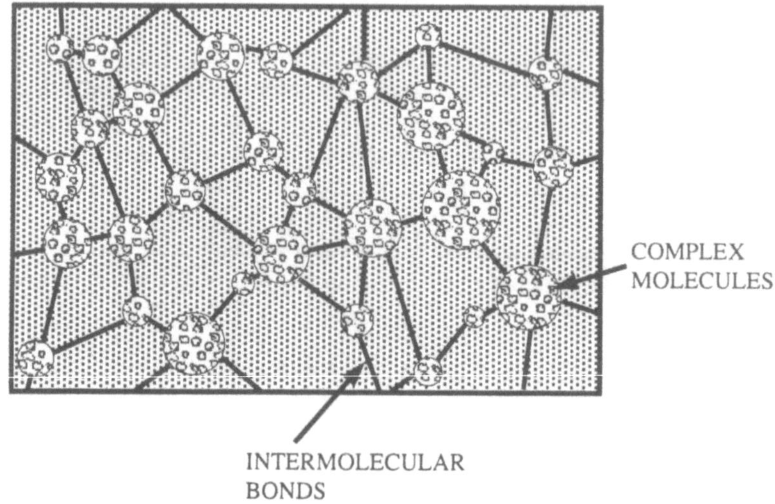

Figure 4.9. An idealized view of intermolecular matrix of homogeneous, amorphous medium.

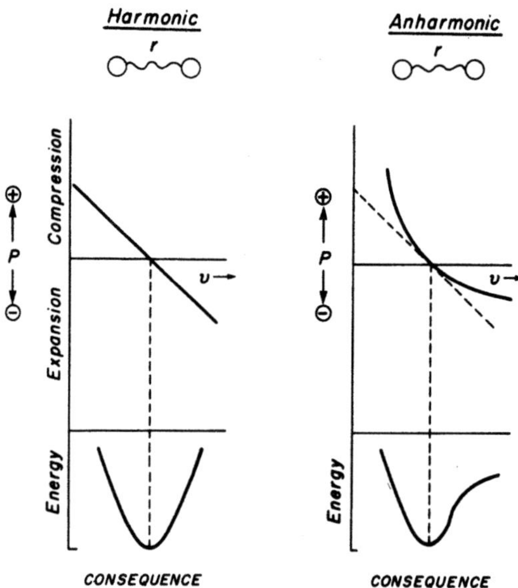

Figure 4.10. Potential functions for harmonic or anharmonic motion within a molecular matrix.

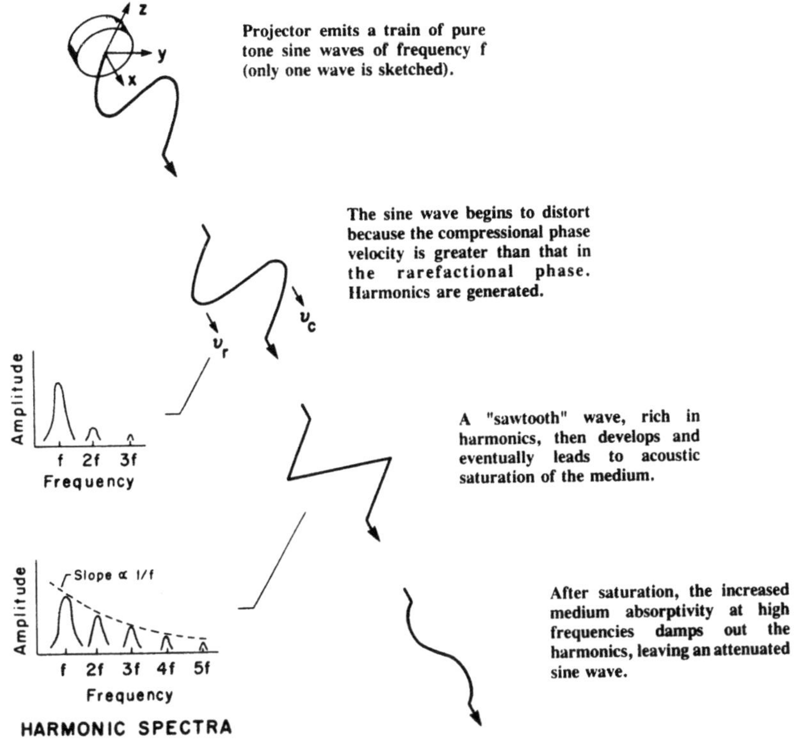

Figure 4.11.   Distortion of ultrasonic pulse as it propagates through a nonlinear medium. (Reprinted with permission from Pergamon Press [22])

$$\beta_S = -\frac{1}{V}\left(\frac{\partial V}{\partial p}\right)_S \tag{4.34}$$

This implies that sound speed is independent of pressure (compare Eqs. [4.33] and [4.34]). If, on the other hand, the oscillator responds in a fashion that makes further compression more difficult, as is illustrated in the right panel of the above figure, the slope of pressure versus volume increases with pressure. This in turn implies that reciprocal-compressibility and, hence, the sound speed will increase with pressure. Thus, different points of a waveform of a propagating wave travel at slightly different speeds. Over an appropriate distance of propagation, the waveform is distorted and eventually leads to shock formation. The sequence of events that leads to shock formation is illustrated in Figure 4.11.

The nonlinear response of a medium to sound propagation is often expressed quantitatively by the variable B/A, which is the ratio of the coefficients of the quadratic to the linear terms of a Taylor series representing the equation of the

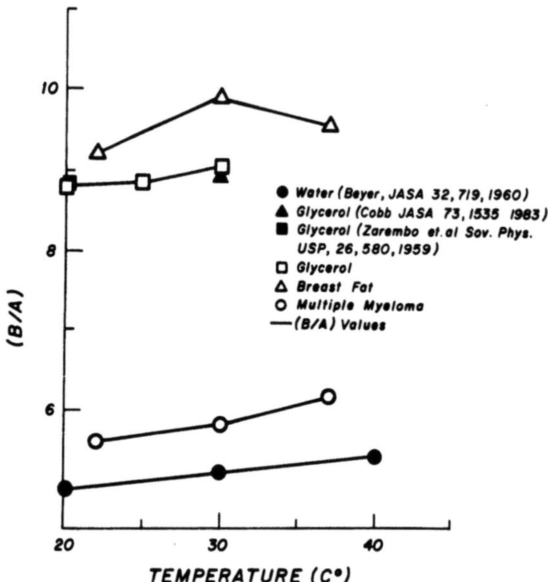

Figure 4.12. Comparison of B/A values at different temperatures.   (Reprinted with permission from Acoustical Society of America [32])

state of the medium. This ratio is related to other acoustic and thermodynamic values by the relationship [23],

$$\frac{B}{A} = 2\rho_0 c_0 \left(\frac{\partial c}{\partial p}\right)_S. \tag{4.35}$$

Because B and A are the coefficients of the equation of state, they represent some fundamental properties of the medium [24]. Various researchers have proposed that the ratio of these coefficients could serve as an index to characterize tissues properties.

A nonlinear phenomenon manifests itself in many different ways. Each of the manifestations can be used to measure or image the B/A parameter. Methods used to measure B/A include measurement of change in phase with pressure [25, 26] and measurement of higher harmonics as a function of distance [27]. Ichida et al. [28] suggested an imaging method involving measurement of the change in phase by low-frequency pump waves. More recently, Cain [29] suggested imaging schemes in reflection mode. Finally, the formation of the sum and difference of two transmitted frequencies due to nonlinearity of a medium has been used to make images [30, 31] .

Figure 4.12 summarizes acoustic measurements of different types of tissues. Figure 4.12 shows B/A values for various tissues. Multiple myeloma, represented by open circles, is a malignant tissue and contains a large fraction of water. Its B/A value ranges between 5.5 and 6 over the temperature range

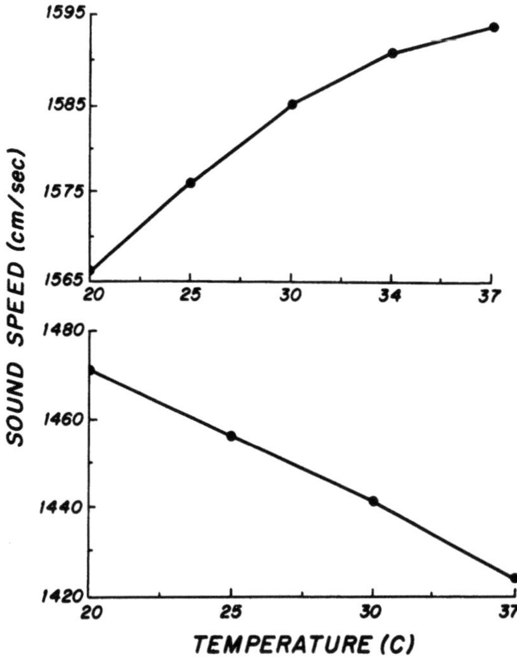

Figure 4.13.   Example illustrating different temperature dependence for soft tissues rich in water content (liver) (top panel) and lipids (subcutaneous fat) (bottom panel).

of 20° C to 37° C. On the other hand, B/A values for breast fat, represented by open triangles, are twice as nonlinear. Also, tissues containing large amounts of fat and water show a marked difference in sound speed.

Figure 4.13 shows sound speed in excised liver containing 72% water by weight. In this case sound speed increases with temperature. In contrast to liver, in subcutaneous fat containing 95% fat, sound speed goes down as temperature is increased. At 37° C there is a difference of 180 m/sec between the two types of tissues. The experimental results of the last two figures imply that there is a close correlation between tissue composition and acoustic properties, i.e., B/A and c [32, 33]. Alternatively, from the measurement of B/A and speed, tissue composition can be determined. Measurements have been made on a wide variety of biologic tissues. For example, often diseased livers exhibit lower sound speed than their normal counterparts. Due to different temperature coefficients, the sound speed differences between normal and abnormal livers are in general more pronounced at higher temperatures. The markedly fatty livers respond differently to temperature variations. They show a decrease in speed with temperature as opposed to an increase observed in normal livers.

The values of B/A for normal livers are about 30% higher than that of water. The samples of cirrhotic and tumorous livers (high water content) show lower values than the normal counterparts. However, fatty tissues exhibit a significantly higher value of nonlinearity.

Assuming biologic tissues to be ideal mixtures of water, fat, and a residual component of proteins, carbohydrates, etc., one can arrive at mixture laws for acoustic parameters. Two approaches, one based on linear addition of compressibility of the constituents [33] and the other based on linear addition of pulse travel time through the components [34], have been proposed. For the sake of simplicity we shall discuss the latter. Compressibility based mixture laws and comparison of the two models are discussed in references [33] and [35], respectively. As an example, consider a mixture that contains fat and proteins uniformly and isotropically distributed in water. A sound pulse is transmitted through the mixture and received at the other end after a propagation distance, L. If the pulse travels a distance d_f and d_p through fat and proteins and if no time delay is caused by the presence of interfaces, the total time T_t taken to travel through the medium is the sum of the times, st_f, t_p, and t_w taken to travel through each component, i.e.,

$$T_t = \sum_{i=f,p,w} t_i \qquad (4.36)$$

or

$$\frac{1}{c} = \frac{d_f}{L}\frac{1}{c_f} + \frac{d_p}{L}\frac{1}{c_p} + \frac{L - d_f - d_p}{L}\frac{1}{c_w}. \qquad (4.37)$$

Assuming that the distance traveled by a pulse is proportional to its volume,

$$\frac{1}{c} = \sum_{i=f,p,w} x_i \frac{1}{c_i}, \qquad (4.38)$$

where x_i represents the volume fraction of the ith component. Differentiating Eq. (4.38) with respect to pressure under the simplifying assumption that volume fractions do not change appreciably with pressure and comparing the results with Eq. (4.35), one obtains

$$N = \sum_{i=f,p,w} x_i N_i, \qquad (4.39)$$

where

$$N = \frac{B/A}{\rho c^3}. \qquad (4.40)$$

It should be noted that

$$\sum_{i=f,p,w} x_i = 1. \qquad (4.41)$$

The mixture laws defined by Eq. (4.38) and 4.39 can be simplified in the following form.

$$\frac{1}{c} = A_0 + A_1 x_w + A_2 x_f \tag{4.42}$$

and

$$N = B_0 + B_1 x_w + B_2 x_f, \tag{4.43}$$

here the constants A and B are related to sound speed, acoustic nonlinearity, and density of the pure components. The subscripts w and f represent water and fat components. Measurements made on excised tissues were fitted to Eqs. (4.42) and (4.43) to determine the constants A and B. It is important to know that it is the reciprocal of sound speed (and not the sound speed) that is related to the volume fraction of individual components. Secondly, magnitude of any of the two acoustic parameters alone does not uniquely identify the tissue composition. This is illustrated by the graphic representation of Eq. (4.42) in Figure 4.14.

First of all one observes some "valleys" in the contour of the three-dimensional plot. These are largely due to the limited set of data used to obtain the above equations. More importantly, for a given value of c, multiple values of water-fat composition are possible. The diagonal lines in the lower panel represent a fixed value of sound speed. Therefore, it is clear that use of any one variable alone to determine tissue composition could be misleading. However, when Eqs. (4.42) and (4.43) are used together and solved as simultaneous equations, one can determine gross composition of tissues. A comparison of the values predicted by Eqs. (4.42) and (4.43) and the actual values is shown in Figure 4.15. There is good agreement between the two sets as is illustrated by unit slope and zero intercept. The analysis proposed by Apfel [33] involves solving mixture laws of density, compressibility, and B/A as simultaneous equations with the aid of a Lagrange multiplier. This method also yields good estimates of tissue composition.

4.2.3 Conclusions

Measurements of sound speed and nonlinearity offer a means of quantitative measurement of major components of tissues and, hence, characterize their state—normal versus pathologic. However, these values cannot be easily measured with required accuracy in *in vivo* or imaging environments. Currently, several attempts are being made to achieve this.

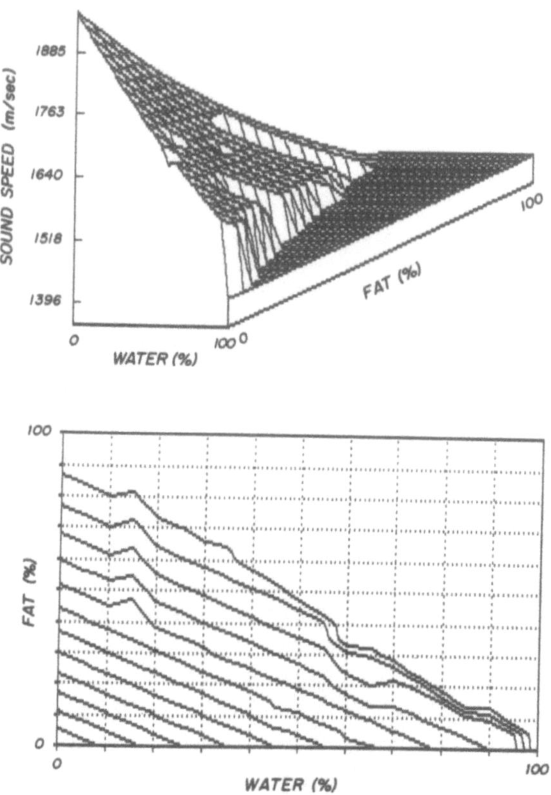

Figure 4.14.   Top panel: Three-dimensional plot of sound speed as a function of water and fat content of tissues. The data for the plot were obtained from Eq. (4.42). The constants A_0, A_1, and A_2 were determined by fitting experimental data to Eq. (4.42). Bottom panel: " Iso-speed " plot as a function of composition. The diagonal lines across the axes correspond to a fixed sound speed.

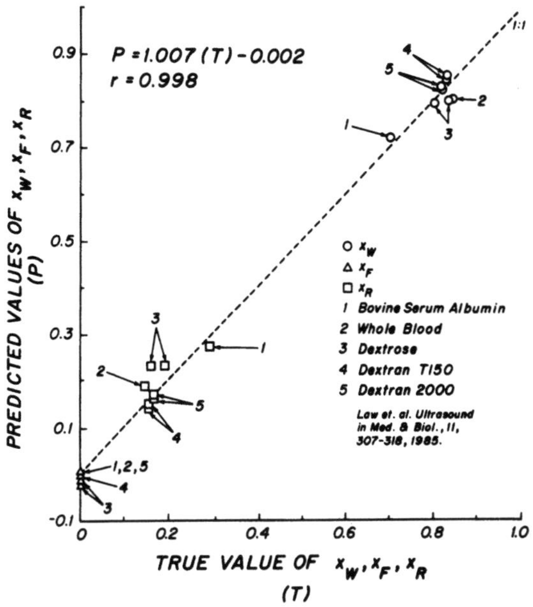

Figure 4.15.   Comparison between the composition predicted by Eqs. (4.42) and (4.43) from the values of B/A and c and the actual composition of the medium reported in [36].

Bibliography

[1]   K. J. Parker, T. A. Tuthill, and R. B. Baggs, "Ultrasound attenuation of glycogen: In vitro and in vivo results," *Proceedings IEEE Ultrasonics Symposium*, vol. 2, pp. 993–996, 1987.

[2]   E. L. Carstensen, "Absorption of sound in tissues," in *Ultrasonic Tissue Characterization II* (M. Linzer, ed.), pp. 29–36, National Bureau of Standards (Special publication no. 525), Washington, DC, 1979.

[3]   F. W. Kremkau, R. W. Barnes, and C. P. McGraw, "Ultrasonic attenuation and propagation speed in normal human brain," *Journal of the Acoustical Society of America*, vol. 70, pp. 29–38, 1981.

[4]   F. Dunn and J. K. Brady, "Absorption of ultrasound in biological media," *Biofizika*, vol. 18, no. 2, pp. 1063–1066, 1973.

[5]   L. W. Kessler, W. D. O'Brien Jr., and F. Dunn, "Ultrasonic absorption in aqueous solutions of polyethylene glycol," *Journal of Physical Chemistry*, vol. 74, pp. 4096–4102, 1970.

[6]   H. Pauly and H. P. Schwan, "Mechanism of absorption of ultrasound in liver tissue," *Journal of the Acoustical Society of America*, vol. 50, pp. 692–699, 1971.

[7] L. J. Slutsky, L. Madsen, R. D. White, and J. Harkness, "Kinetics of the exchange of protons between hydrogen phosphate ions and histidyl residue," *Journal of Physical Chemisty*, vol. 84, pp. 1325–1329, 1980.

[8] S. A. Goss, L. A. Frizzell, and F. Dunn, "Ultrasonic absorption and attenuation in mammalian tissues," *Ultrasound in Medicine and Biology*, vol. 5, pp. 181–186, 1979.

[9] F. W. Kremkau, "Biomolecular absorption of ultrasound. III. Solvent interactions," *Journal of the Acoustical Society of America*, vol. 83, pp. 2410–2415, June, 1988.

[10] D. R. Raichel, "Sound propagation in Voigt fluid," *Journal of the Acoustical Society of America*, vol. 52, pp. 395–398, 1972.

[11] A. S. Ahuja, "Ultrasonic attenuation in soft tissues: Reasons for large magnitude and linear frequency dependence," *Ultrasonic Imaging*, vol. 2, pp. 391–396, October, 1980.

[12] J. D. Pohlhammer, C. A. Edwards, and W. D. O'Brien, Jr., "Phase insensitive ultrasonic attenuation coefficient determination of fresh bovine liver over an extended frequency range," *Medical Physics*, vol. 8, pp. 692–694, 1981.

[13] M. O'Donnell and J. G. Miller, "Mechanisms of ultrasonic attenuation in soft tissue," in *Ultrasonic Tissue Characterization II* (M. Linzer, ed.), pp. 37–40, National Bureau of Standards (Special publication no. 525), Washington, DC, 1979.

[14] F. Dunn, P. D. Edmonds, and W. J. Fry, "Absorption and dispersion of ultrasound in biological media," in *Biological Engineering* (H. P. Schwan, ed.), pp. 205–332, McGraw-Hill Book Co., 1969.

[15] J. C. Bamber, "Attenuation and absorption," in *Physical Principles of Medical Ultrasonics* (C. R. Hill, ed.), pp. 118–199, Ellis Horwood Limited. Chichester, England, 1986.

[16] D. Shore and C. A. Miles, "Attenuation of ultrasound in homogenates of bovine skeletal muscle and other tissues," *Ultrasonics*, vol. 26, pp. 218–223, July, 1988.

[17] M. O. Woods and C. A. Miles, "Ultrasound speed and attenuation in homogenates of bovine skeletal muscle," *Ultrasonics*, vol. 24, pp. 260–266, September, 1986.

[18] C. M. Sehgal and J. F. Greenleaf, "Ultrasonic absorption and dispersion in biological media: A postulated model," *Journal of the Acoustical Society of America*, vol. 72, pp. 1711–1718, 1982.

[19] S. Leeman, "Ultrasound pulse propagation in dispersive media," *Physics in Medicine and Biology*, vol. 25, pp. 481–488, 1980.

[20] R. M. Arthur and K. V. Gurumurthy, "A single-pole model for the propagation of ultrasound in soft tissues," *Journal of the Acoustical Society of America*, vol. 77, pp. 1589–1597, 1985.

[21] H. A. H. Jongen, J. M. Thijssen, M. van den Aarssen, and W. A. Verhoef, "A general model for the absorption of ultrasound by biological tissues and experimental verification," *Journal of the Acoustical Society of America*, vol. 79, pp. 535-540, 1986.

[22] T. G. Muir and E. L. Carstensen, "Prediction of nonlinear acoustic effects at biomedical frequencies and intensities," *Ultrasound in Medicine and Biology*, vol. 6, pp. 345-357, 1980.

[23] R. T. Beyer, "Parameter of nonlinearity in fluids," *Journal of the Acoustical Society of America*, vol. 32, pp. 719-721, 1960.

[24] B. Hartmann, "Potential energy effects on the sound speed in liquids," *Journal of the Acoustical Society of America*, vol. 65, pp. 1392-1396, 1979.

[25] Z. Zhu, M. S. Roos, W. N. Cobb, and K. Jensen, "Determination of the acoustic nonlinearity parameter B/A from phase measurements," *Journal of the Acoustical Society of America*, vol. 74, pp. 1518-1521, 1983.

[26] C. M. Sehgal and J. F. Greenleaf, "Scattering of ultrasound by tissues," *Ultrasonic Imaging*, vol. 6, pp. 60-80, January, 1984.

[27] F. Dunn, W. K. Law, and L. A. Frizzell, "Nonlinear ultrasonic propagation in biological media," *British Journal of Cancer*, vol. 45(Suppl. 5), pp. 55-58, 1982.

[28] N. Ichida, T. Sato, and M. Linzer, "Imaging the nonlinear ultrasonic parameter of a medium," *Ultrasonic Imaging*, vol. 5, pp. 295-299, October, 1983.

[29] C. A. Cain, "Ultrasonic reflection mode imaging of the nonlinear parameter B/A: I. A theoretical basis," *Journal of the Acoustical Society of America*, vol. 80, pp. 28-32, 1986.

[30] C. M. Sehgal, B. Porter, and J. F. Greenleaf, "Relationship between acoustic nonlinearity and the bound and the unbound states of water," *IEEE Ultrasonic Symposium*, vol. 2, pp. 883-886, 1985.

[31] Y. Nakagawa, M. Nakagawa, M. Yoneyama, and M. Kikuchi, "New nonlinear parameter imaging CT system using a parametric acoustic array," in *Acoustical Imaging* (A. J. Berkhout, J. Ridder, and L. F. van der Wal, eds.), vol. 4, pp. 595-604, Plenum Press, New York, 1985.

[32] C. M. Sehgal, R. C. Bahn, and J. F. Greenleaf, "Measurement of the acoustic nonlinearity parameter B/A in human tissues by a thermodynamic method," *Journal of the Acoustical Society of America*, vol. 76, pp. 1023-1029, 1984.

[33] A. F. Apfel, "Prediction of tissue composition from ultrasonic measurements and mixture rules," *Journal of the Acoustical Society of America*, vol. 79, pp. 148-152, 1986.

[34] C. M. Sehgal, G. M. Brown, R. C. Bahn, and J. F. Greenleaf, "Measurement and use of acoustic nonlinearity and sound speed to estimate composition of excised livers," *Ultrasound in Medicine and Biology*, vol. 12, pp. 865-874, November, 1986.

[35] E. C. Everbach, "Tissue composition determination via measurement of the acoustic nonlinearity parameter," *Thesis (Yale University)*, New Haven, CT 1989.

[36] W. K. Law, L. A. Frizzell, and F. Dunn, "Determination of the nonlinearity parameter B/A of biological media," *Ultrasound in Medicine and Biology*, vol. 11, pp. 307–318, 1985.

5
Class 1, 2, and 3 Scattering

5.1 Introduction

Scattering processes have been most successful in imaging and characterization of biologic tissues. In the traditional sense scattering may be defined as the change of *amplitude*, *frequency*, *phase velocity*, or *direction of propagation* of a wave as a result of spatial or temporal nonuniformities of the medium. The inhomogeneities arise because of variations in compressibility or acoustic impedance differences. In this section, we almost exclusively treat scattering as a phenomenon of redirection of energy in space. A few techniques that are currently used to extract quantitative information about the nature of the tissues in the echo mode are discussed.

The scattering behavior of biologic tissues like liver, heart, and kidney is complex and depends on several factors listed below.

1. The magnitude of acoustic impedance or compressibility difference between the scatterer and the surrounding medium. Larger differences produce stronger scattering, all other factors being equal.

2. The spatial distribution of inhomogeneities, whether periodic, random, or intermediate between the two limits. Such scattering can have long-range as well as short-range order.

3. Scale of inhomogeneities. The term "scale" often implies the size of scatterers and the interspacing between them. It is generally defined by a single variable a'. If a' is significantly less than λ, there is Rayleigh scattering; if a' is comparable to λ, diffraction occurs; and finally, if a' is very large compared to λ, the object behaves like a polished surface and geometric phenomena like reflection and refraction dominate. Although all three categories constitute tissues, the second category—diffractive scatterers—has been studied most widely to elucidate information on the properties of tissues. Propagation through such scatterers is no longer along rays. Instead, the energy transmission and scattering behavior can be best explained in terms of wave fronts.

4. Changes in the number of scatterers per unit volume also influence the echogenicity of the medium. Most theoretical treatments to explain scattering in tissues use the Born approximation to solve the wave equation. The rationale for this assumption is that the measured scattering cross sections are small. The implication is that multiple scattering effects can be ignored. This may not always be true. The higher-order effects could play an important role 1) in explaining the frequency dependence of energy "loss" due to scattering [1] and 2) in the application of the Fourier-slice theorem (which assumes weak scattering) for diffraction tomography [2]. However, from the point of view of our present discussion, we shall assume the Born approximation to be valid for all practical purposes.

5.2 Model-Based Scatter Analysis

5.2.1 Signal Analysis

Backscattering (i.e., scattering toward the transmitter) is the most widely used phenomenon for ultrasonic imaging and tissue characterization. Two approaches are currently being used. They involve signal processing of either raw RF data or envelope-detected data. Each of the methods has some advantages. Generally speaking, RF signal analysis is more straightforward and simpler to implement. On the other hand, envelope-detected signals can be obtained and stored with less sampling and memory [3, 4]. It is the envelope-detected signal that is displayed in the conventional ultrasound B-scan images.

5.2.2 RF Analysis

Consider a transducer arrangement shown in Figure 5.1. A transducer with radius a is located at a distance R from a tissue volume V. The length L and the lateral dimension of this volume are given by the range gating function and the transducer directivity function, respectively.

In an inhomogeneous medium the inhomogeneities arise due to spatial *fluctuation of* density ($\Delta\rho$) and speed (Δc) about the mean value ρ_0 and c_0. The wave equation for sound propagation in such an inhomogeneous medium is [6]

$$\frac{1}{c^2}\left(\frac{d^2p}{dt^2}\right) - \nabla^2 p + \nabla log(\rho \nabla p) = 0, \tag{5.1}$$

where p is the acoustic pressure. For a weakly scattering system,

$$\frac{\Delta\rho}{\rho_0} << 1, \; and \; \frac{\Delta c}{c_0} << 1. \tag{5.2}$$

By using a small perturbation method Eq. (5.1) is solved for the scattered pressure p_s. Following this, the pressure p_s is averaged over aperture area A of the transducer to give backscatter pressure p_m [5],

$$p_m = (i2kp_0A)^{-1} \int_V \left[2\frac{\Delta c}{c_0}k^2 p_I^2 + \nabla\left(\frac{\Delta\rho}{\rho_0}\right)p_I \nabla p_I \right] idV. \tag{5.3}$$

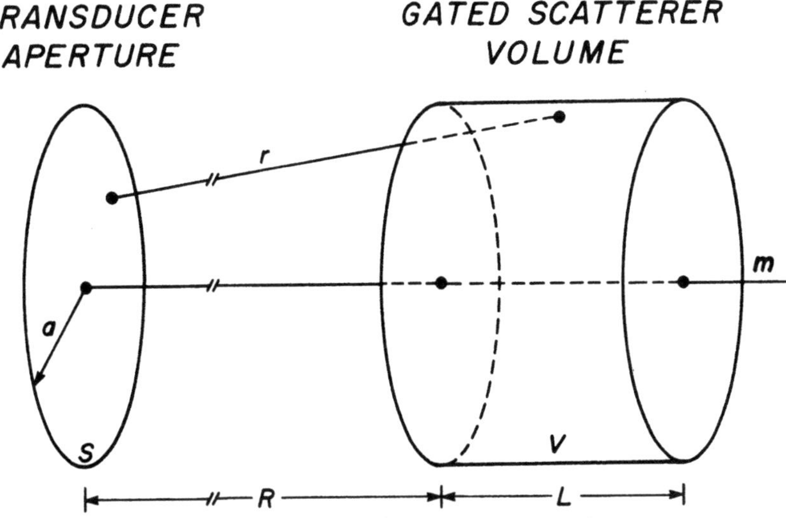

Figure 5.1.   Geometry for analysis of tissue scattering observed by a transducer centered along scan line m. (Reprinted with permission from Acoustical Society of America [5].)

The subscript m represents the mth scan line of a B-scan image. Eq. (5.3) represents two types of scattering: due to velocity fluctuation and due to density variation. In addition to this, the equation also contains the term p_I, which is the incident pressure. This implies that the measured pressure is influenced not only by the fundamental properties of tissues, i.e., $\Delta\rho/\rho_0$ and $\Delta c/c_0$, but also by the structure of the incident field. Because the aim of tissue characterization is to identify tissue properties, it is necessary to remove the influence of the incident field. Several approaches have been suggested in the literature. As an illustration consider the approach described by Lizzi et. al. [5].

For a weakly focused field of the type shown in Figure 5.2, p_I is defined as

$$p_I\left(R^{'},\theta\right) = ikAp_0F(\theta)\left(\frac{e^{-ikR^{'}}}{2\pi R'}\right). \qquad (5.4)$$

$F(\theta)$ and p_0 in the above equation represent the directivity function and the amplitude of vibration at the transducer, respectively. Substitution of the expression for p_I in that for p_m followed by Fourier transformation and simplification yields the measured spectrum of the received signal $v_m(\omega)$.

$$v_m(\omega) = H(\omega)k^2Ap_0(\omega)\left(\frac{e^{-i2kR}}{2(2\pi R)^2}\right) \qquad (5.5)$$
$$\int\int\int T(x,y,z)F^2(\omega,y,z)g(x)e^{-i2kx}dxdydz,$$

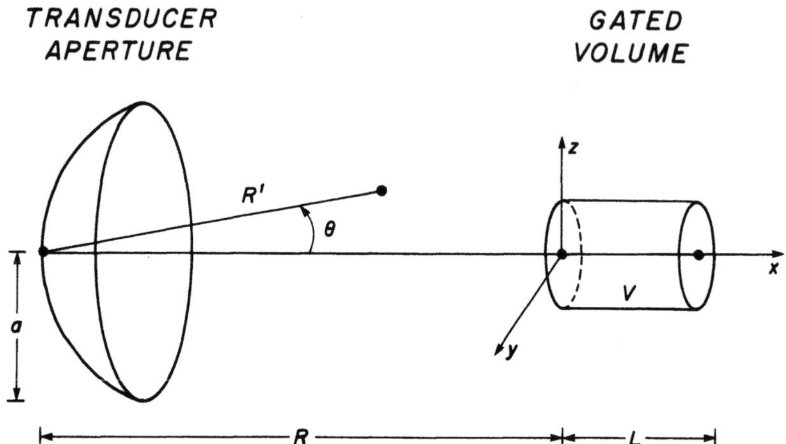

Figure 5.2.   Geometry for the analysis of tissue scattering with focused transducers. (Reprinted with permission from Acoustical Society of America [5].)

where $H(\omega)$ is the linear transfer function for the transducer and associated electronics; $p_0(\omega)$ is the spectrum of the radiated pulse; $F^2(\omega, y, z)$ is a two-way directivity function of the transducer and is described by Bessel functions; $g(x)$ is the gating function; and $T(x, y, z)$ is the rate of change of impedance (Z) along x, i.e.,

$$T(x, y, z) = \frac{1}{Z_0} \frac{\partial Z}{\partial x}, \tag{5.6}$$

where Z_0 is the local average of acoustic impedance.

Several factors that relate to the incident field and the transducer characteristics affect the measured spectrum $v_m(\omega)$. These factors are removed by dividing the spectrum of a signal with that of the signal reflected from a flat glass plate. The normalized spectrum $s_m(\omega)$ is expressed by

$$s_m(\omega) = \int \left[\int \int T(x, y, z) D^2(y, z) \right] g(x) e^{-i2kx} dx, \tag{5.7}$$

where $D^2(y, z)$ is the normalized directivity function. The power spectrum averaged over M scan lines of the region of interest is given by

$$S(\omega) = \frac{1}{M} \sum_{m=1}^{M} \left| s_m(\omega)^2 \right|, \tag{5.8}$$

or

$$S(\omega) = 4k^2 \int \left[\int \int R_\zeta(\Delta x) R_D(\Delta x, \Delta z) d(\Delta y) d(\Delta z) \right] \cdot$$
$$R_G(\Delta x) e^{-i2k\Delta x} d(\Delta x), \tag{5.9}$$

where R_ζ, R_D, and R_G represent autocorrelation functions of relative impedance, directivity function, and gating function, respectively. In Eq. (5.9), R_D, and R_G are known. By using an appropriate autocorrelation function R_ζ and the measured $S(\omega)$, one can derive the tissue properties. For example, for Rayleigh scatterers the particle concentration forms a Poisson process [7] and the autocorrelation function is

$$R_\zeta(\Delta x) = u\delta(\Delta x) + u^2, \qquad (5.10)$$

where u is mean particle concentration and δ is the Dirac delta function. Substitution of Eq. (5.9) in Eq. (5.10) and using appropriate definitions of R_G and R_ζ, the power spectrum is described as [5]

$$S(\omega) \simeq \left\{ 0.36u\frac{a^2}{\pi R^2}\left[\int g^2(x)k^4 dx\right]\right\}k^4 + 4u^2\left[\int g(x)e^{i2kx}dx\right]^2 k^2. \quad (5.11)$$

The bracketed portion of the second term on the right-hand side is a Fourier transform of the slowly varying function $g(x)$. It does not contribute at the high frequency of interest and, therefore, can be neglected. Hence,

$$S(\omega) \approx K^* u f^4, \qquad (5.12)$$

where

$$K^* = 0.36\frac{a^2}{\pi R^2}\left[\int g^2(x)dx\right] \qquad (5.13)$$

or

$$log[S(\omega)] = 4log(f) + log(K^* u). \qquad (5.14)$$

Thus, from the slope and intercept of Eq. (5.14) one can determine the frequency dependence (and hence an estimate of size of the particle) and the relative particle concentration, respectively. Figure 5.3 summarizes the preceding analysis scheme for asteroid hyalosis. The frequency relationship shown in the above figure approximates f^4, thus implying Rayleigh scattering. Over the measured frequency band, this would correspond to scatterers with diameters less than 35 μm.

Fellingham and Sommer [8] and Sommer et al. [9] also analyzed RF signals. They studied tissues with structural organization on the order of the ultrasonic wavelength. They treated the tissue as an isotropic, semiregular array of scatterers with a definable mean spacing d.

For scatterers separated by a distance d along the propagation path, the backscattered signal from adjacent scatterers will differ in phase (ϕ) by $k(2d)$. For $\phi = n\pi$, i.e., $n\lambda = 2d$, there will be constructive and destructive interference

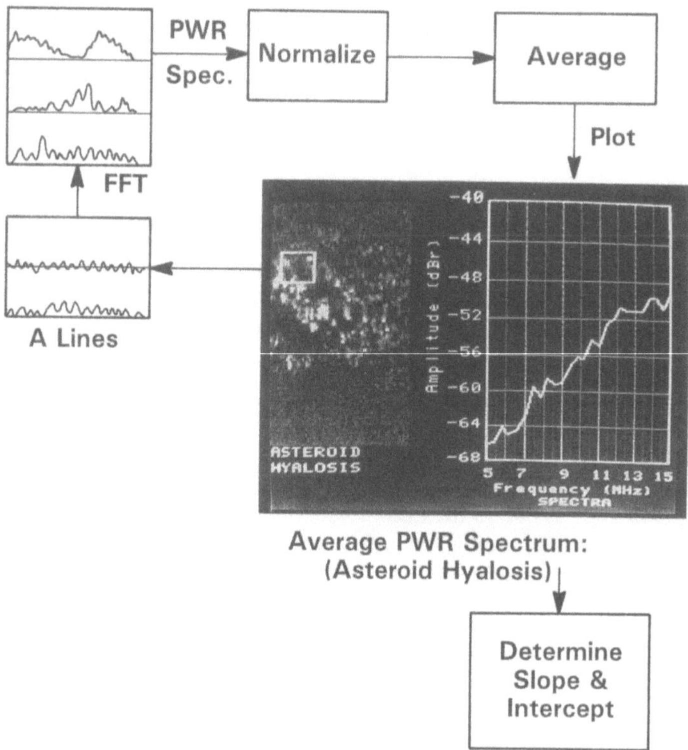

Figure 5.3. Scheme of power (PWR) spectrum analysis of the marked region within asteroid hyalosis. The A lines from the selected region are Fourier transformed. The power spectrum of each line is normalized by the power spectrum of the signal reflected by a flat glass plate. The power spectrum is averaged and the information obtained from the slope and intercept is used to characterize the properties of tissues. (Reprinted with permission from Acoustical Society of America [5].)

for n equal to integer and integer divided by 2, respectively. If the medium is insonated by a broad band pulse, a complex interference pattern will result. That is, the frequency spectrum of the backscatter signal will show peaks at frequencies $f = nc/2d$ or the difference between the periodic peaks will at

$$\Delta f = \frac{c}{2d}. \tag{5.15}$$

Figure 5.4 shows an example of backscatter signal and its spectrum. It is clear from the data that it is difficult to determine the periodicity. To evaluate periodicity due to acoustic interference, the magnitudes of various spectra were

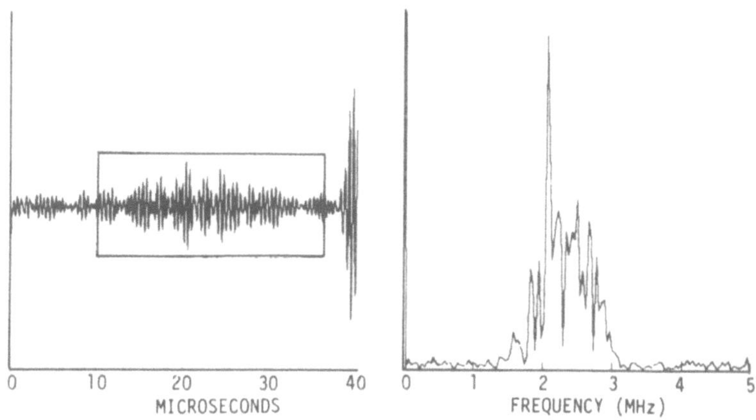

Figure 5.4. Typical backscatter signal and its Fourier spectrum. (From [10])

autocorrelated and the first peak was used to yield Δf. These in turn were used
to calculate d by using the last equation. Measurements made on several scan
lines yielded ranges of scatterer spacings in the tissue. The various steps are
summarized in Figure 5.5.

The mean scatterer spacings in normal and abnormal livers as determined
by the above procedures show (Table 5.1) that average scatterer spacings are
measurably different for tissues with different pathology, thus offering a potential
for characterizing tissues by ultrasound.

5.2.3 Statistical Analysis of Envelope-Detected Signal

Wagner and co-workers [11–13] developed methods of characterizing tissues
by using an envelope (or its square, i.e., intensity) of complex RF signals.
Their work shows how the RF signals get mixed together when the envelope
or intensity signal is used and to what extent the mixing can be undone to
reveal tissue properties. This group of investigators initiated the scattering
class idea used in this book. Three of the five classes of scatterers (1, 2,
and 3) may be separable by their techniques. Class 1 scatterers are present
in sufficient concentration to contribute to the echo signal. These scatterers
are distributed randomly in space and give rise to a signal that follows circular
Gaussian statistics. Because it is assumed that the positions of these scatterers
are uncorrelated, they give rise to incoherent scattering, I_d. Class 2 scatterers are
nonrandomly distributed, with long-range order defined by an average spacing d.
These scatterers lead to coherent or specular scattering with mean intensity I_s and
variance $var(I_s)$. These scatterers appear to be very similar to those determined
by Fellingham and Sommer [8, 9] and described earlier. Class 3 scatterers are

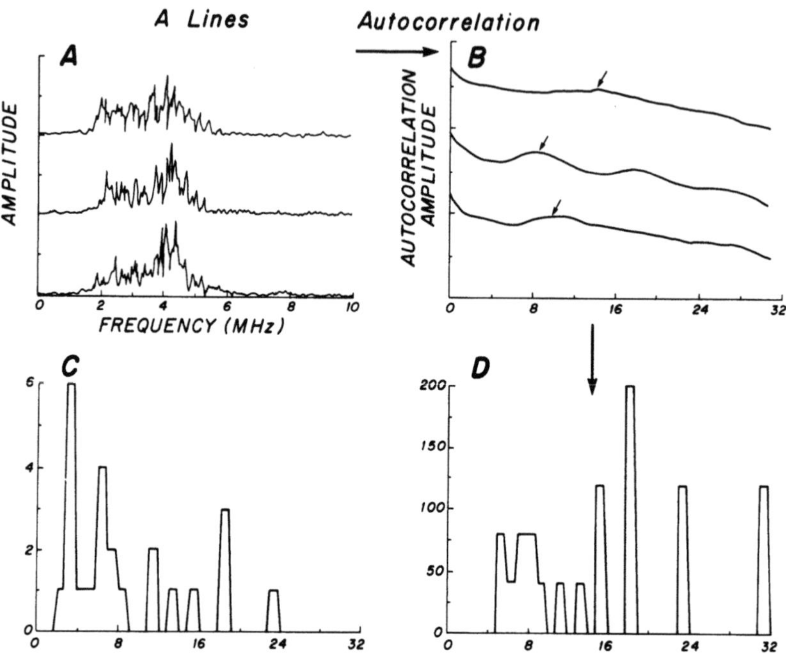

Figure 5.5.   (A) Frequency spectra of A lines of ultrasonic image of normal liver.
(B) Autocorrelation of frequency spectra.   Arrows represent computer-detected first
autocorrelation peaks.   (C) and (D) scatter spacings for normal and cirrhotic livers,
respectively.   (Reprinted with permission from The Radiological Society of North
America [9])

Table 5.1   Scatterer spacing in normal and abnormal livers. n represents the number
of cases.[1]

	Mean spacing (SD), mm
Normal subjects (n = 14)	1.07 (0.16)
Cirrhosis (n = 15)	1.48 (0.24)
Chronic active hepatitis (n = 2)	1.16 (1.22)

[1](Reprinted from L. L. Fellingham and F. G. Sommer, "Ultrasonic characterization of tissue
structure in the in vivo human liver and spleen," IEEE Trans. Sonics Ultrasonics, vol. SU-31, pp.
418-428, July, 1984 with permission from The Institute of Electrical and Electronics Engineers
© 1984 IEEE [8])

structures like organ surfaces or blood vessels. These scatterers are nonrandom with short-range order and require non-Gaussian methods for statistical analysis. Such structures are eliminated before tissue analysis on images is done.

If the complex ultrasonic echo signal (along the transducer beam axis) from diffuse and specular classes of scatterers is represented by

$$g_d = g_R + i g_I \tag{5.16}$$

and

$$h_s = h_R + i h_I, \tag{5.17}$$

respectively, the net signal is

$$g = (g_R + h_R) + i(g_I + h_I) \tag{5.18}$$

or intensity is

$$I = |g|^2. \tag{5.19}$$

We are interested in the correlation between intensity values at two positions z_1 and z_2 along the z (or the transducer) axis, i.e.,

$$R(z_1, z_2) = |I_1 I_2| = |g_1|^2 |g_2|^2, \tag{5.20}$$

where $R(z_1, z_2)$ represents the autocorrelation function. The above equation has been further developed in the form [11, 12]

$$R(z_1, z_2) = R(\Delta z) =$$
$$I_d^2(1 + |\rho^*|^2 + 2I_d I_s + <I_{s1} * I_{s2}> + 2I_d\rho <h_{R1} * h_{R2}>), \tag{5.21}$$

where ρ_* is the complex coherence factor and $<*>$ means average lagged product. The correlation function described by Eq. (5.21) is solved for three cases so as to obtain variables that can be directly measured from ultrasonic images.

Case 1: $\Delta z = z_2 - z_1 = 0$

$$t \equiv R_I(0) = 2I_d^2 + 4I_d^2 I_s + I_s^2. \tag{5.22}$$

Case 2: $\Delta z = nd$, where n is an integer,

$$p \equiv R_I(nd) = I_d^2 + 2I_d I_s + I_s^2. \tag{5.23}$$

Case 3: $\Delta z \gg d$

$$b \equiv R_I(\Delta z \gg d) = (I_d + I_s)^2. \tag{5.24}$$

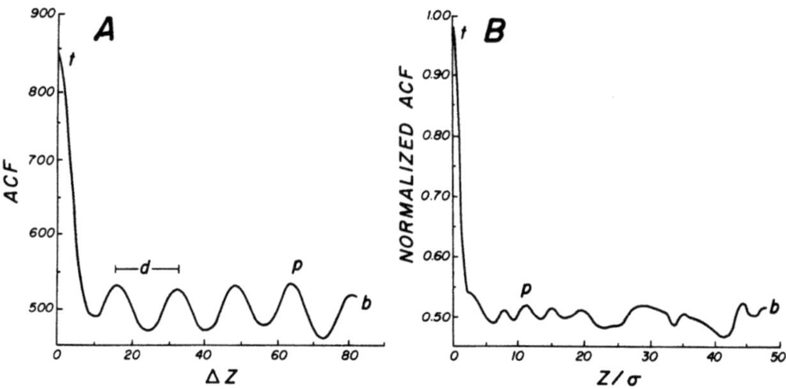

Figure 5.6.   (A) Autocorrelation function (ACF) of intensity for simulated Rician speckle with single frequency structured specularity of period d. (Reprinted with permission from Society of Photo-Optical Instrumentation Engineers [11]) (B) Average autocorrelation function in the range of direction for an intensity image of a normal human liver. Region of interest ~2 cm^2. (Reprinted with permission from Society of Photo-Optical Instrumentation Engineers [12])

In the above equations t, p, and b represent top, peak, and bottom of the autocorrelation function shown in Figure 5.6. The parameters d, p, and b, although easily identifiable in an idealized case (as in Figure 5.6A) cannot be easily determined for experimental data. The problem is overcome by solving Eqs. (5.22) to (5.24) as simultaneous equations.

$$I_d = \sqrt{b} - \sqrt{b - t + p} \tag{5.25}$$

$$I_s = \sqrt{b - t + p} \tag{5.26}$$

$$var^{\frac{1}{2}}(I_s) = \sqrt{p - b} \tag{5.27}$$

The parameters b and t are determined by taking the square of the first moment and second moment of the intensity image, respectively. The measurement of p is more difficult because for most soft tissues, correlation peaks are small compared to uncertainty in the measurement. The problem is overcome by using an alternative approach.

(a) Taking the power spectrum of the intensity image. This is shown for normal human liver in Figure 5.7.

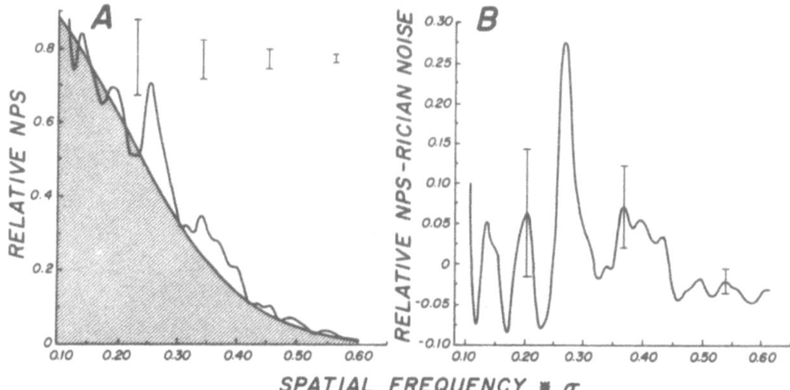

Figure 5.7. (A) Average noise power spectrum (NPS) in intensity image of a normal human liver. The shaded area is the estimate of the Rician noise contribution to the intensity variation. (B) Power spectrum of (A) after Rician noise has been subtracted. (Reprinted with permission from Society of Photo-Optical Instrumentation Engineers [12])

(b) Estimating the Rician noise contribution. This is achieved by fitting the spectral minima of the spectrum to a Gaussian function.

(c) Subtracting the Rician noise, shown as shaded area in the figure, from the original power spectrum.

(d) Integrating the residual spectrum shown in the right panel of Figure 5.7.

It has been shown theoretically that the integral over frequency is the variance in I_s which in turn is equal to $(p - b)$ by Eq. (5.27).

Once the values of t, b, and p are known, they are used in the last three equations to determine the ratio, r, of specular to diffuse scattering and the fractional standard deviation in the specular intensity.

$$r = \frac{I_s}{I_d} \tag{5.28}$$

$$v = var^{1/2}(I_s)/I_s \tag{5.29}$$

Also, the average scatter spacing d is determined from the average frequency of incommensurate spectral peaks of the power spectrum. The three parameters r, v, and d are used to describe tissue architecture and its microstructure. Measurements of these backscatter parameters from normal volunteers are tabulated in Table 5.2.

Although no one characteristic uniquely defines each organ, in combination these values can serve as sensitive indices to characterize tissues. As more

Table 5.2 Measurements of scattering in some organs in vivo.

Organ	Patients, no.	d (SD), mm	r (SD)	v (SD)
Kidney	7	1.32 (0.23)	0.78 (0.20)	3.03 (0.90)
Spleen	7	1.30 (0.10)	0.31 (0.08)	3.19 (1.60)
Liver	-	1.13 (0.10) ($n = 22$)	0.23 (0.13) ($n = 8$)	2.62 (0.74) ($n = 8$)

data accumulate, it may be possible in the future to identify tissue type and pathology by clustering [14].

5.3 Texture-Based Scatter Analysis

5.3.1 Classification

There is a category of ultrasound tissue characterization methods that does not fit in the classification scheme based on the nature of scatterers. These methods are heuristic in nature and deal with the recognition of patterns of texture of the ultrasonic video images. Pattern recognition can mean various things depending on the context in which it is used. Typically, it is defined as a process in which an observer (human or computer) is shown a selected number of samples of different classes and then an object or image of unknown class and is required to categorize it in one of the classes. The term "texture" as used here does not have a rigid description either. It can be regarded as a measure of properties like fineness, coarseness, smoothness, granulation, randomness, irregularity, or inhomogeneity [15]. These properties are generally described either in terms of structural elements and the "grammar" by which they repeat spatially or by computing indices or parameters that quantitatively describe the stochastic properties of spatial distributions of gray levels in an image. Various researchers have applied the second category of methods to characterize ultrasonic images [16–28].

Textural patterns in ultrasonograms are determined not only by tissue microstructure but also by the properties of the imaging system [29]. Also, axial and lateral textures of the ultrasonogram should be considered separately because they are influenced by unrelated phenomena, i.e., spectral content of RF signal, transducer geometry, time-gain compensation and sample positions [29]. Therefore, when using computerized analysis to determine texture, technical specifications of the clinical system and the experimental details of the gain-compression curve and the time-gain compensation must be taken into consideration in analyzing the results. In other words, pattern recognition techniques developed for analyzing various other forms of images [15], in principle, should not be directly applied to ultrasonograms unless the asymmetry of axial and lateral spread functions and the effects of focusing on lateral resolution are accounted for. Although some of these issues could profoundly affect the outcome of an experimental study, a large number of studies have ignored these aspects. A wide variety of parameters has been chosen to characterize ultrasonic images. Some of the methods that have been used on clinical data are described below.

5.3.2 Gray-Level Run-Length Texture Analysis

A large proportion of clinical studies in this area have used run-length statistics to differentiate spatial patterns of echo amplitudes of tissues. A gray-level run-length is defined as number of consecutive pixels that have a similar gray-level value in a given direction of an image. This involves formation of a run-length matrix from the image matrix. This process is illustrated by the example described in [22]. If a patch of digitized image pixels is represented by

$$
\begin{matrix}
0 & 1 & 2 & 3 \\
0 & 2 & 3 & 3 \\
2 & 1 & 1 & 1 \\
3 & 0 & 3 & 0
\end{matrix}
\tag{5.30}
$$

its run length matrices in horizontal and vertical directions will be given by

$$
\begin{array}{c|cccc}
0^0 \; RL & 1 & 2 & 3 & 4 \\
\hline
GL & & & & \\
0 & 4 & 0 & 0 & 0 \\
1 & 1 & 0 & 1 & 0 \\
2 & 3 & 0 & 0 & 0 \\
3 & 3 & 1 & 0 & 0
\end{array}
\tag{5.31}
$$

$$
\begin{array}{c|cccc}
90^0 \; RL & 1 & 2 & 3 & 4 \\
\hline
GL & & & & \\
0 & 2 & 1 & 0 & 0 \\
1 & 4 & 0 & 0 & 0 \\
2 & 3 & 0 & 0 & 0 \\
3 & 3 & 1 & 0 & 0,
\end{array}
\tag{5.32}
$$

respectively.

Generally, before the run-length matrix is computed the image data are preprocessed. The digitized images are first normalized so that all the images have a comparable uniform gray level frequency distribution. Also the gray-level and the run lengths are grouped into different ranges. For example, a normalized image (32x32) with gray level ranging from 0 to 127 is grouped into four ranges: 0-31, 32-63, 64-95, and 96-128. Similarly, the run lengths are grouped into ranges: 1-3, 4-7, 8-15, and 16-31, to give a 4x4 array from which run-length parameters are computed. Although in principle any direction can be chosen, typically for ultrasound images run-length matrices are computed in horizontal (azimuthal) and vertical (axial) directions. If $p(i,j)$ represents the i, j^{th} element of the run-length matrix, where i and j represent the gray-level and run-length values, respectively, the run-length statistics are defined by several parameters defined by the following equations:

$$LRE = \frac{\sum_{i=1}^{p} \sum_{j=1}^{q} j^2 p(i,j)}{\sum_{i=1}^{p} \sum_{j=1}^{q} p(i,j)} \qquad (5.33)$$

$$SRE = \frac{\sum_{i=1}^{p} \sum_{j=1}^{q} \frac{p(i,j)}{j^2}}{\sum_{i=1}^{p} \sum_{j=1}^{q} p(i,j)}, \qquad (5.34)$$

$$GLN = \frac{\sum_{i=1}^{p} \left[\sum_{j=1}^{q} p(i,j) \right]^2}{\sum_{i=1}^{p} \sum_{j=1}^{q} p(i,j)}, \qquad (5.35)$$

$$RLN = \frac{\sum_{j=1}^{q} \left[\sum_{i=1}^{p} p(i,j) \right]^2}{\sum_{i=1}^{p} \sum_{j=1}^{q} p(i,j)}, \qquad (5.36)$$

$$RP = \sum_{i=1}^{p} \sum_{j=1}^{q} \frac{p(i,j)}{N_{roi}}. \qquad (5.37)$$

In the above equations, LRE, SRE, GLN, RLN, and RP represent the parameters long-run emphasis, short-run emphasis, gray-level nonuniformity, run-length nonuniformity, and run percentage, respectively. The variables p and q represent the number of gray-level and run-length groups, respectively. N_{roi} is the number of pixels in the region of interest. The above parameters can be described in terms of visual parameters like "coarseness" or "fineness" of an image [30]. Long runs of adjacent pixels imply coarse texture, whereas the magnitude of short runs is indicative of fineness or high-frequency content of texture. Gray-level nonuniformity is smaller for coarser images and increases as the image texture increases in fineness.

By using run-length analysis it was shown that the spatial pattern of echocardiographic images differentiates amyloid and hypertrophic cardiomyopathy from normal myocardium [31]. Also, run-length parameters and Markovian statistics on ultrasound B-scan images, when used in conjunction with mammography, aided the separation of cysts and solids and improved the false-positive rate of mammographic diagnosis by approximately 40% of the studied cases [28].

5.3.3 Markovian Statistics for Texture Analysis

Another set of measurements that has been used successfully to analyze texture of biomedical images is based on Markovian statistics [32]. The essence of the simplest form of a Markov process is that the outcome of the nth trial is dependent only on the $(n-1)^{th}$ trial. That is, the trials are neither statistically independent nor depend on any other trial preceding the $(n-1)^{th}$ trial. This analysis is based on synthesizing a matrix of gray-level transition probabilities $P_L(i/j)$ from the gray values of digital images. Transition probability $P_L(i/j)$ is defined as the conditional probability of gray level i occurring L pixels after gray level j occurs. To illustrate, let us consider a 4x4 image represented by the matrix

$$
\begin{matrix}
0 & 1 & 2 & 3 \\
1 & 3 & 1 & 2 \\
2 & 0 & 3 & 1 \\
1 & 3 & 3 & 3.
\end{matrix}
\tag{5.38}
$$

For simplification, if one considers a jump (or transition) of only one pixel at a time (L=1), the transition probability matrix will be described by the matrix

$$
\begin{matrix}
i & 0 & 1 & 2 & 3 \\
j & & & & \\
0 & 0 & 1 & 0 & 0 \\
1 & 0 & 1 & 2 & 2 \\
2 & 0 & 0 & 1 & 1 \\
3 & 0 & 3 & 0 & 3
\end{matrix}
\tag{5.39}
$$

divided by the total number of jumps, which in the present example is 10. Such matrices are calculated for different step sizes L. From each such matrix, several Markov texture parameters are computed [32] to describe image texture. In general, one can regard run length and Markovian parameters as complementary to one another: whereas run length describes elements of consistency in an image, the Markovian parameters describe elements of variation.

Of the various measured parameters, the uncorrelated parameters are used to classify images into different groups. Systematic application of these parameters to ultrasound images has not been made. If one takes the results obtained from digital microscopic images as guidance, it is clear that the selection of step size strongly affects the degree of classification error. The percentage of correct classification is maximum when Markov step size is near or below the limit of the scanning system. Reduction in step size below this value does not change the classification performance significantly, because redundant information is available at these small step sizes. On the other hand, high resolution is progressively lost with increasing step sizes. The lost information results in deterioration of the performance of the measured variables.

Bibliography

[1] C. M. Sehgal and J. F. Greenleaf, "Scattering of ultrasound by tissues," *Ultrasonic Imaging*, vol. 6, pp. 60–80, January, 1984.

[2] M. Azimi and A. C. Kak, "Distortion in diffraction tomography caused by multiple scattering," *IEEE Transactions on Medical Imaging*, vol. MI-2, pp. 176–195, December, 1983.

[3] P. He and J. F. Greenleaf, "Application of stochastic analysis to ultrasonic echoes–estimation of attenuation and tissue heterogeneity from peaks of echo envelope," *Journal of the Acoustical Society of America*, vol. 79, pp. 526–534, February, 1986.

[4] P. He and J. F. Greenleaf, "Attenuation estimation on phantoms–a stability test," *Ultrasonic Imaging*, vol. 8, pp. 1–10, 1986.

[5] F. L. Lizzi, M. Greenebaum, E. J. Feleppa, M. Elbaum, and D. J. Coleman, "Theoretical framework for spectrum analysis in ultrasonic tissue characterization," *Journal of the Acoustical Society of America*, vol. 73, pp. 1366–1373, 1983.

[6] L. A. Chernov, *Wave Propagation in a Random Medium*. New York: Dover Publications, 1960.

[7] F. L. Lizzi and M. E. Elbaum, "Clinical spectrum analysis techniques for tissue characterization," in *Ultrasonic Tissue Characterization II* (M. Linzer, ed.), pp. 111–119, National Bureau of Standards (Special publication no. 525), Washington, DC, 1979.

[8] L. L. Fellingham and F. G. Sommer, "Ultrasonic characterization of tissue structure in the in vivo human liver and spleen," *IEEE Transactions on Sonics and Ultrasonics*, vol. SU-31, pp. 418–428, July, 1984.

[9] F. G. Sommer, L. F. Joynt, B. A. Carroll, and A. Macovski, "Ultrasonic characterization of abdominal tissues via digital analysis of backscattered waveforms," *Radiology*, vol. 141, pp. 811–817, 1981.

[10] L. Joynt, D. Boyle, H. Rakowski, R. Popp, and W. Beaver, "Identification of tissue parameters by digital processing of real-time ultrasonic clinical cardiac data," in *Ultrasonic Tissue Characterization II* (M. Linzer, ed.), pp. 267–273, National Bureau of Standards (Special publication no. 525), Washington, DC, 1979.

[11] R. F. Wagner, M. F. Insana, and D. G. Brown, "Unified approach to the detection and classification of speckle texture in diagnostic ultrasound," *Optical Engineering*, vol. 25, pp. 738–742, June, 1986.

[12] M. F. Insana, R. F. Wagner, B. S. Garra, D. G. Brown, and T. H. Shawker, "Analysis of ultrasound image texture via generalized Rician statistics," *Optical Engineering*, vol. 25, pp. 743–748, June, 1986.

[13] R. F. Wagner, M. F. Insana, and D. G. Brown, "Statistical properties of radio-frequency and envelope-detected signals with applications to medical ultrasound," *Journal of Optical Society of America [A]*, vol. 4, pp. 910–922, 1987.

[14] A. Chu, C. M. Sehgal, and J. F. Greenleaf, "Use of gray value distribution of run lengths for texture analysis," *Pattern Recognition Letters*, vol. 11, pp. 415–419, 1990.

[15] R. M. Haralick, "Statistical and structural approaches to texture," *Proceedings of the IEEE*, vol. 67, pp. 786–804, 1979.

[16] J. M. Thijssen and A. M. Verbeek, "Computer analysis of A-mode video echograms from choroidal melanoma," in *Ultrasonography in Ophthalmology* (J. M. Thijssen and A. M. Verbeek, eds.), vol. 8, pp. 123–139, Dr. W. Junk Publishers. The Hague, 1981.

[17] D. Nicholas, A. Barrett, J. M. G. Chu, D. O. Cosgrove, P. Garbutt, J. Green, S. Pussell, and C. R. Hill, "Computer analysis of gray scale tomograms," in *Acoustical Imaging* (A. F. Metherell, ed.), vol. 8, pp. 731–744, Plenum Press, New York, 1980.

[18] R. A. Lerski, E. Barnett, P. Morley, P. R. Mills, G. Watkinson, and R. N. M. MacSween, "Computer analysis of ultrasonic signals in diffuse liver disease," *Ultrasound in Medicine and Biology*, vol. 5, pp. 341–350, 1979.

[19] R. A. Lerski, M. J. Smith, P. Morley, E. Barnett, P. R. Mills, G. Watkinson, and R. N. M. MacSween, "Discriminant analysis of ultrasonic texture data in diffuse alcoholic disease. 1. Fatty liver and cirrhosis," *Ultrasonic Imaging*, vol. 3, pp. 164–172, April, 1981.

[20] O. R. Mitchell, C. R. Myers, and W. Boyne, "A max-min measure for image texture analysis," *IEEE Transactions on Communications*, vol. C26, pp. 408–414, 1977.

[21] M. S. Good, J. L. Rose, and B. Goldberg, "Application of pattern recognition techniques to breast cancer detection: ultrasonic analysis of 100 pathologically confirmed tissue areas," *Ultrasonic Imaging*, vol. 4, pp. 378–396, October, 1982.

[22] M. M. Galloway, "Texture analysis using gray level run lengths," *Computer Graphics and Image Processing*, vol. 4, pp. 172–179, 1975.

[23] A. K. Bhandari and N. C. Nanda, "Two-dimensional echocardiographic recognition of abnormal changes in the myocardium," *Ultrasound in Medicine and Biology*, vol. 8, pp. 663–671, 1982.

[24] D. J. Skorton, H. E. Melton, Jr., N. G. Pandian, J. Nichols, S. Koyanagi, M. L. Marcus, S. M. Collins, and R. E. Kerber, "Detection of acute myocardial infarction in closed-chest dogs by analysis of regional two-dimensional echocardiographic gray-level distributions," *Circulation Research*, vol. 52, pp. 36–44, 1983.

[25] D. J. Skorton, S. M. Collins, J. Nichols, N. G. Pandian, J. A. Bean, and R. E. Kerber, "Quantitative texture analysis in two-dimensional echocardiography: application to the diagnosis of experimental myocardial contusion," *Circulation*, vol. 68, pp. 217–223, 1983.

[26] J. F. Greenleaf, K. Chandrasekaran, and H. A. McCann, "Advanced tissue analysis and display with ultrasound," *Proceedings of Xth Brazilian Conference on Biomedical Engineering*, vol. 4, no. 2, pp. 49–60, 1987.

[27] K. Chandrasekaran, R. C. Bansal, J. F. Greenleaf, A. Hauck, J. B. Seward, A. J. Tajik, and L. L. Bailey, "Early recognition of heart transplant rejection by backscatter analysis from serial 2D echos in a heterotopic transplant model," *Journal of Heart Transplantation*, vol. 6, pp. 1–7, 1987.

[28] V. Goldberg (Murmis), J. J. Gisvold, T. M. Kinter, and J. F. Greenleaf, "Texture analysis of ultrasound B-scans to aid diagnosis of cancerous lesions in the breast," *IEEE Ultrasonics Symposium*, vol. 2, pp. 839–842, 1988.

[29] S. W. Flax, G. H. Glover, and N. J. Pelc, "Textural variations in B-mode ultrasonography: a stochastic model," *Ultrasonic Imaging*, vol. 3, pp. 235–257, July, 1981.

[30] D. J. Skorton, S. M. Collins, and R. E. Kerber, "Digital image processing and analysis in echocardiography," in *Cardiac Imaging and Image Processing* (S. M. Collins and D. J. Skorton, eds.), pp. 171–205, McGraw-Hill Company. New York, 1986.

[31] K. Chandrasekaran, P. E. Aylward, S. R. Fleagle, T. L. Burns, J. B. Seward, A. J. Tajik, S. M. Collins, and D. J. Skorton, "Feasibility of identifying amyloid and hypertrophic cardiomyopathy with the use of computerized quantitative texture analysis of clinical echocardiographic data," *Journal of the American College of Cardiology*, vol. 13, pp. 832–840, 1989.

[32] N. J. Pressman, "Markovian analysis of cervical cell images," *Journal of Histochemistry and Cytochemistry*, vol. 24, pp. 138–144, 1976.

6
Class 4 Scattering

6.1 Introduction

So far we have dealt with a special class of scatterers that are static in nature. In living systems, tissues and organs are in constant motion to accomplish their biologic functions. Change in normal blood flow or the movements associated with heart, lung, and gut borders are often indicative of the physical state of the organ. As the tissue scatterers move during ultrasonic irradiation, one observes a scattering behavior that is typical of motion. For example, the frequency of reflected sound is shifted as the reflector moves with respect to the transducer; the displacement of a reflecting surface like that of a heart valve modulates the echo signal about a mean position; or if a scattering medium is interrogated with two pulses separated by an interval, the pulses are temporally shifted closer or away due to the movement of scatterers. Any of these phenomena like frequency shift, echo modulation, or temporal shift between the two pulses during round-trip travel to a scatterer can, in principle, be used to obtain information on motion. Techniques based on these phenomena are discussed below. Each of the techniques compensates for the weakness of the other and in turn suffers from some limitation of its own.

6.2 M-Mode Ultrasound

The principle of motion mode, more commonly known as M-mode, can be described with the aid of Figure 6.1 in which a reflector is moving up and down about a mean position. The echo from the reflector will arrive either early or late as the reflector moves toward or away from the transducer, respectively. If the echo amplitude, represented by a bright spot on a cathode-ray tube screen, is swept across at a constant rate, a wavy trace of the type shown on the right of Figure 6.1 is obtained. For a static scatterer, the trace will be a horizontal line.

Examples of the use of M-mode in echocardiography to interrogate a sagittal view of heart along a long-axis plane in three different directions are shown in Figure 6.2. The y-axis are the distance from the transducer and the x-axis is time. The waveforms represent displacements as a function of time. From the comparison it is clear that different parts of the organ execute a complex and variable motion. Recognition of patterns of the waveforms often forms the basis of diagnosis.

M-mode has been used extensively in evaluating the motion of heart walls or heart valves. The majority of studies involve qualitative recognition of traces.

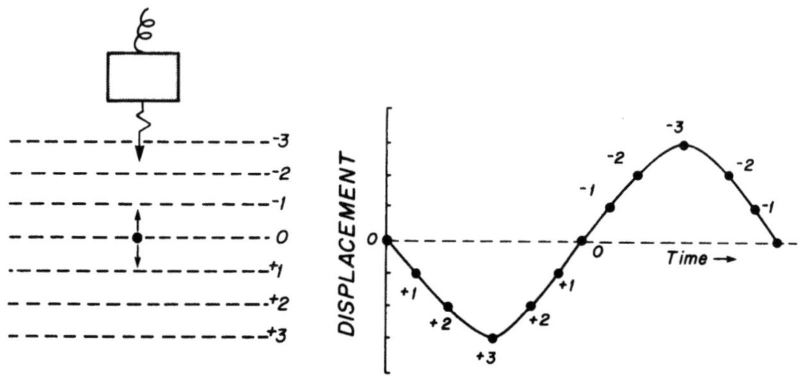

Figure 6.1. Echoes from a moving surface are swept across a screen, producing a time history of the position of scatterers in the stationary beam.

There is now a growing interest in using M-mode digitized images to obtain quantitative information regarding the nature of the tissue. In a recent study it was shown by using cross-correlation techniques on digitized, time-averaged M-mode images that cardiac motion in fetuses deforms the neighboring lung, and that these deformations can be determined quantitatively and related to the physiologic state of the organ [2].

6.3 Doppler Methods

Major applications of real-time B-mode scanners have been to image tissue structures in a narrow slice of insonated region. Although this methodology has been successful in various areas of diagnosis in medicine, it suffers from some limitations that can potentially lead to interpretation errors. For example, when an artery to be imaged is curved and scanned tangentially, the image can suggest stenosis even though it may not be present. Thrombus and plaques rich in lipids having ultrasonic properties not significantly different from blood are often difficult to detect. On the other hand, calcified plaques may attenuate so strongly that they can cast strong shadows that obscure other parts of an atheroma and vessel walls. Loss of information that may arise due to such weaknesses can be compensated for by observing blood flow through the vessels directly. Also, blood circulation through the capillary network transports materials needed for the proper function of the organ. Obstruction in the blood supply affects the performance of the organ. Therefore, measuring blood flow through vessels has been of great interest and value in medicine. Several imaging methods are currently used to acquire this information. Doppler methods provide hemodynamic information directly and noninvasively. As the name suggests, these methods make use of the Doppler effect (Fig. 6.3), according to which, if ultrasound of known frequency, f_0, is targeted onto a moving scatterer,

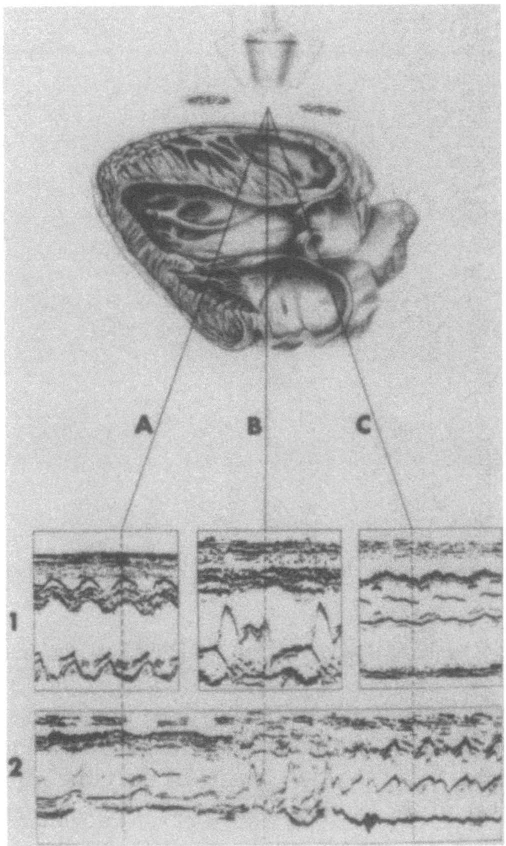

Figure 6.2.   Motions of the wall of the heart as depicted by M-mode imaging. (Reprinted with permission from Churchill Livingstone [1])

the frequency, f_r, of the signal reflected toward the transmitter is shifted in proportion to the velocity, v, of the reflector, according to the equation

$$f_D = f_r - f_0 = \frac{2f_0 v \cos\theta}{c}, \tag{6.1}$$

where θ is the angle between the interrogating beam and the direction of blood flow and c is the speed of sound in the medium.

It is mainly the erythrocytes in blood that scatter ultrasound. These cells thus act as markers for measuring the blood flow velocity. It is important to note that Doppler methods measure blood velocity and not volumetric flow rate. The latter, an important physiologic variable, is computed from the knowledge of blood velocity and its profile and the cross-sectional area of lumen. In practice, diagnostic frequencies of 2 MHz to 10 MHz are used. At these frequencies,

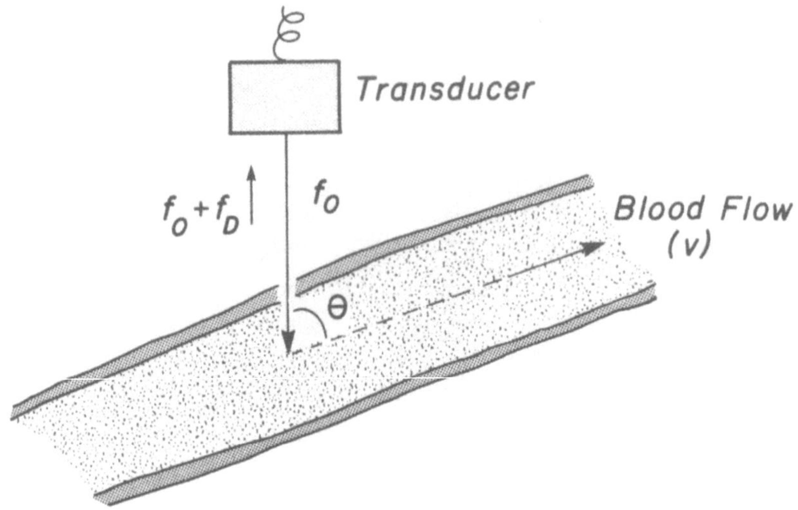

Figure 6.3. Doppler shift caused by insonation of a moving scatterer such as moving blood.

motion of biologic systems leads to Doppler shifts in the audible frequency range.

Several Doppler techniques are currently used in medicine. Each of these techniques has settled into its diagnostic niche. Often they are used in conjunction, each compensating for the shortcoming of the other approaches. These techniques have been in use for several years and have been described in detail in several excellent texts [3, 4].

6.3.1 Various Techniques

Continuous-Wave Doppler

The simplest Doppler instrument uses continuous waves of ultrasound to detect motion. Such an instrument uses two crystals: one emitting sound waves continuously and the other acting as a receiver of the backscattered frequency-shifted signal. The detected signal is amplified and demodulated to extract the Doppler-shifted frequency. Different methods of demodulation have been used. The simplest one mixes the echo signal with the reference RF signal to produce beat frequencies equal to the Doppler frequencies. The mixed signal is rectified, low-pass filtered, amplified, and used to drive audio speakers. However, this method of demodulation does not distinguish between the flow toward or away from the transducer. To obtain directional information, heterodyne or phase quadrature demodulators are used. The latter provides two audio signals which are sometimes detected with stereo headphones and represent two opposing directions of flow.

In practice, the backscatter signal is obtained from many scatterers or many surfaces moving at different speeds. The result is a spectrum of audio frequencies that change in time. Generally, frequencies below 100 Hz, that arise due to slowly moving vessels or breathing, are removed by a filter placed before the audio amplifier. Because this modality deals with measuring frequency shifts, it is important that the transmitted signal is as narrow a band as possible. This is achieved by using high-Q transducer material and by air backing the transducer.

Continuous-wave Doppler units are often used to detect movement of the fetal heart. These instruments have broad beams so that small changes in the position of the fetus do not cause a significant loss in signal.

Pulsed Doppler

Continuous–wave Doppler lacks depth resolution. Because the medium is continuously irradiated, it is not known where the reflector is in the beam. To overcome this problem, the medium is insonated by pulses at a constant repetition rate. In between pulses the reflected energy is gated into a receiver. By moving the gate, flow can be measured at different gate positions across the lumen. If the axial length of the sample volume under investigation is ΔX, then the target blood cell will move across the beam in the time Δt described by the equation

$$\Delta t = \frac{\Delta X}{v}. \tag{6.2}$$

A particle moving over this interval will modulate the amplitude of the reflected signal. That is, the reflected signal will be both amplitude and frequency modulated. It is well known that amplitude modulation of a single frequency causes spreading of its frequency spectrum. The frequency spread is inversely proportional to Δt, or

$$\Delta f \propto \frac{v}{\Delta X}. \tag{6.3}$$

Spreading in frequency will lead to uncertainty in velocity Δv. Replacing Δf by the proportional quantity Δv yields

$$\frac{\Delta v}{v} \Delta X = constant. \tag{6.4}$$

The last equation is a reminder of the uncertainty principle. That is, velocity and position cannot be measured precisely simultaneously. For continuous-wave Doppler for which ΔX approaches infinity, $\Delta v \to 0$. This indicates that although the position of echo is not known, the velocity of scatterers can be measured precisely. In contrast to this, as the interrogating pulse is made smaller, $\Delta X \to 0$. This, in essence, means that the measurement of velocity becomes ambiguous. In short, range-measuring capabilities degrade the velocity resolution at the expense of spatial resolution or vice versa. Pulse mode enables localization of echoes, which from a clinical point of view is a useful feature. However, the method is limited in the peak velocities it can measure without aliasing the signal. This is discussed in the next section.

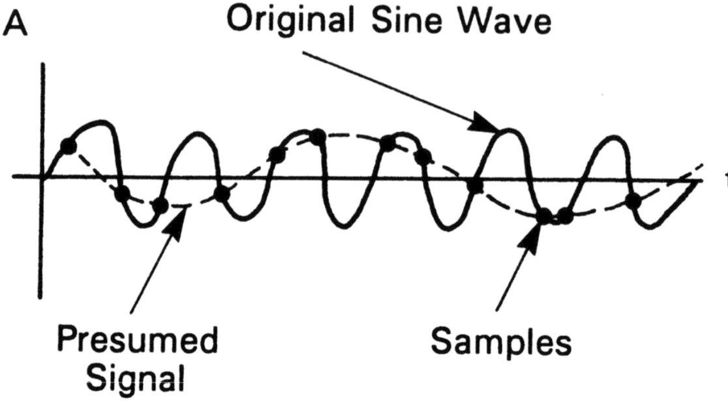

Figure 6.4. Aliasing of Doppler signal due to inadequate sampling.

Aliasing

In the pulsed mode the Doppler signal is sampled at a rate called the pulse-repetition frequency (PRF). To accurately measure frequency shift, the Nyquist theorem must be satisfied. That is, PRF should at least be equal to, or possibly greater than, twice the measured Doppler frequency. This suggests motivation to increase PRF to the highest possible value. Unfortunately, there is an upper limit to which the PRF can be increased. Because sound pulses take finite time to travel to and from the reflectors, one has to wait for some time before the next pulse is launched to avoid overlap between transmitted and reflected signals. Second, PRF should be as low as possible to minimize the exposure of patients to ultrasound. If the blood flow is so large that the Doppler shift is greater than twice the PRF, then there would be inadequate sampling of the signal, resulting in aliasing. This phenomenon is shown in Figure 6.4. The very high portion of the Doppler signal wraps around because of aliasing.

The velocities that can be measured without aliasing can be determined as follows. If the PRF is F_R and the distance of the scatterer from the transmitter is d,

$$F_R \le \frac{c}{2d} \tag{6.5}$$

to avoid interference between the two successive transmitted pulses. That is, a given pulse must complete its round-trip before the next pulse is emitted. Also, to satisfy the sampling theorem without causing aliasing, the maximum Doppler frequency $f_D(max)$ detectable is

$$f_D(max) \le \frac{F_R}{2} \le \frac{c}{4d}. \tag{6.6}$$

The maximum velocity, $v(max)$, measurable is obtained by using the Doppler equation in Eq. (6.6)

$$v(max) \leq \frac{c^2}{8df_0}. \qquad (6.7)$$

Thus pulsed Doppler systems have a velocity cap above which they are unsuitable for measuring velocities. Instruments with high PRF have been designed to overcome this limitation. Such systems are a cross between continuous-wave and pulsed methods and involve transmitting a second pulse before the round-trip of the previous pulse is over. This enables measurement of high velocities but results in range ambiguity. Probably the main advantage of such a system is that it uses a single transducer as opposed to the two used by continuous-wave methods.

Duplex Scanners

Duplex instruments exploit the complementary nature of pulse echo and Doppler systems. Typically in such instruments, a region of interest is marked with a cursor on a B-mode image. Doppler measurements are subsequently made on the marked region of interest. During such an operation, the instrument shares time in the two modes. It is important to note that pulse echo and Doppler systems, although providing complementary information, seem to demand mutually opposing functions from the ultrasound machine. To get better sensitivity or better signal-to-noise ratios for Doppler measurements, the sample volume must be large, which requires transmission of long pulses. On the other hand, the spatial resolution of an echo machine depends inversely on the pulse length. Thus, there is a continuous trade-off between sensitivity and spatial resolution in duplex instruments.

6.3.2 Doppler Tracings

Doppler signals from flowing blood consist of a spectrum of frequencies that arise as a result of complex flow patterns. The velocity of blood varies across the lumen because of viscous friction between the blood and the vessel wall. The blood velocity is near zero at the wall and largest in the center. The profile resembles a parabola. Flow acceleration, for example during pulsatile flow in the aorta, flattens the profile. The flow patterns are further complicated by the shape of the vessel, distance from the inlet, and streaming and turbulence phenomena. Each of these factors affects the Doppler trace. Almost all the modern instruments can display these traces as graphs showing Doppler spectra as a function of time (Fig. 6.5).

Frequency information, as shown in the top panel of Figure 6.5, is divided into discrete bins and displayed along the y-axis (Fig. 6.5 bottom panel). A column of such bins corresponds to the time interval over which frequency analysis is performed. This time interval generally ranges between 1 and 5 ms. The vertical bins are displayed in shades of gray, with brightness

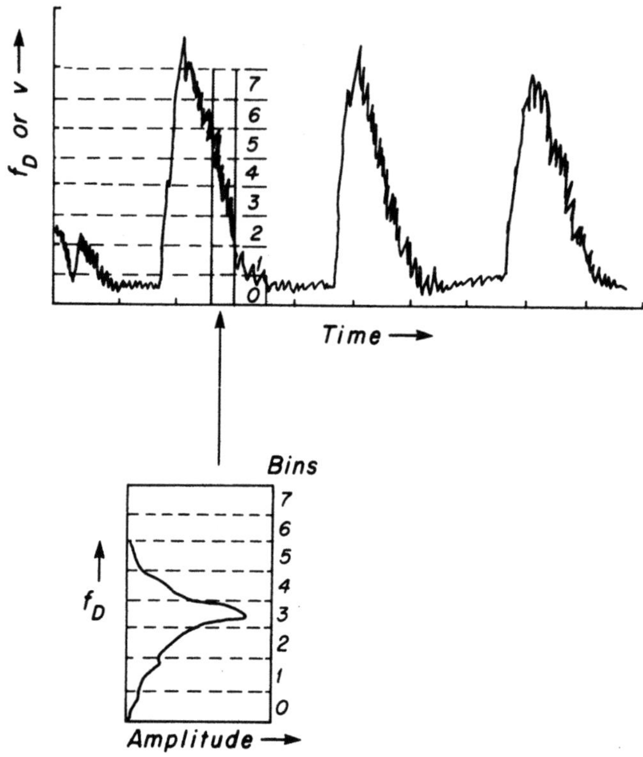

Figure 6.5. Illustration of Doppler tracing. Top panel shows the Doppler spectrum which is segmented in bins. These bins are displayed as a tracing shown in the bottom panel.

proportional to the fraction of signal contained within that frequency range. The stronger the signal, the brighter (darker in this figure) is its representation on the trace. Velocities of erythrocytes traveling toward (away from) the transducer are represented above (below) the baseline. From the characteristics of Doppler spectral displays, diagnostic information like flow, pressure drop across obstructions, and valvular and other circulatory disorders is derived [5].

6.3.3 Doppler Color Flow Mapping

In the last decade, significant advances have been made in real-time display of directional and magnitude information for blood velocity on two-dimensional echographic images. This is achieved by determining the flow direction and the magnitude of each pixel in a sector image. At the scanning rate of 15 to 30 frames/s, such methods require a large number of directional and frequency analyses in a relatively short interval of time. Although "flow direction" measurements can be made rapidly to fulfill the requirements imposed by the time constraints, the frequency measurements by conventional Fourier transform methods are not fast enough.

A method that has been successfully introduced for Doppler imaging is based on correlation methods [6]. Unlike conventional B-scan images which are constructed by transmitting and receiving single pulses along A lines of image frames, Doppler color flow imaging involves transmission and reception of several successive pulses along a given line of sight (LOS) before moving to the next line of the image frame. Echo signals from each transmitted pulse are divided into several contiguous gated regions. Direction of flow and average Doppler shift and its variance are determined from the analyses of signals from each of these gated regions over all the transmitted pulses. These parameters are displayed in color superimposed on B-scan images. Each color pixel of the image corresponds to a gated region of Doppler measurements. The scheme of such an imaging system is conceptualized in Figure 6.6.

Although several pixels of an image are "insonated" along a given LOS, let us consider a sequence of events that occurs in one pixel, i, that lies along the path of interrogation. Echo from region i due to the incidence of first pulse is quadrature detected and used to detect the polarity of the Doppler shift. This information indicates the direction of flow, which is coded as red or blue depending on whether the blood flow is toward or away from the transducer.

The reflections from the subsequent transmitted pulses along a given direction of scanning are used to determine mean Doppler shift and its variance by using an autocorrelator. This method was first proposed by Kasai et al. [7] and is outlined below.

The complex envelope $Z(t)$ of an echo signal is described by

$$Z(t) = X(t) + jY(t), \qquad (6.8)$$

where $X(t)$ and $Y(t)$ are the real and the imaginary parts of the signal. These two components can be directly obtained by using phase quadrature detection. If P(ω) represents the power spectrum of $Z(t)$, the mean frequency of spectrum, $<\omega>$, and its variance, σ^2, can be represented by the first and the second moments, respectively,

$$<\omega> = \frac{\int\limits_{-\infty}^{\infty} \omega P(\omega)d\omega}{\int\limits_{-\infty}^{\infty} P(\omega)d\omega} \qquad (6.9)$$

and

$$\sigma^2 = \frac{\int\limits_{-\infty}^{\infty} (<\omega> - \omega)^2 P(\omega)d\omega}{\int\limits_{-\infty}^{\infty} P(\omega)d\omega}. \qquad (6.10)$$

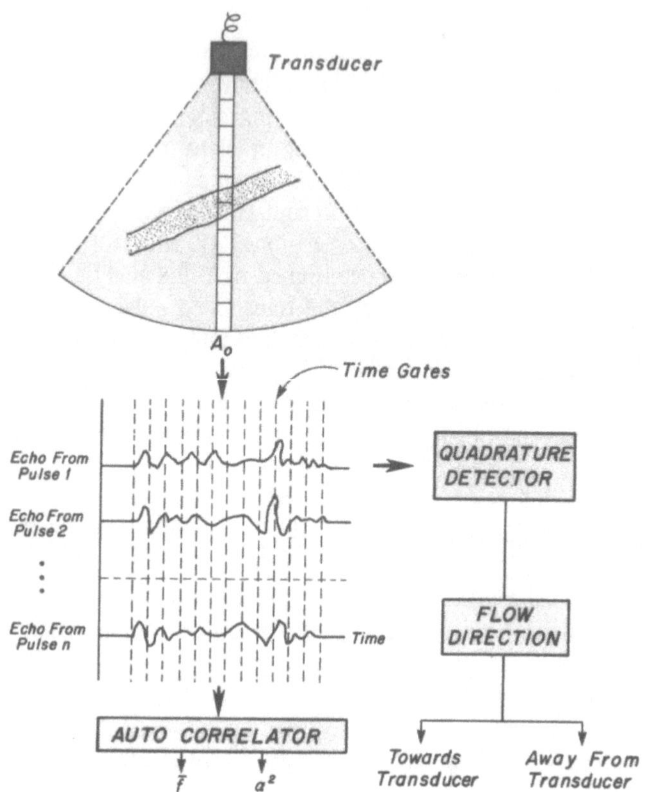

Figure 6.6. Several ultrasound pulses are transmitted along a line of sight (LOS) A_0. Echoes from successive transmitted pulses are received and time gated such that each of the sampled gates corresponds to a pixel along the line of transmission. Echo signal from the reflection of the first transmitted pulse is used to determine flow direction for each pixel. Also, the echoes from the successive transmitted pulses are used with an autocorrelator to determine average Doppler shift and its variance.

According to the Wiener-Khintchine theorem, autocorrelation $R(\tau)$ and the power spectrum are related by the Fourier relationship

$$R(\tau) = \int_{-\infty}^{\infty} P(\omega)e^{j\omega t}d\omega. \tag{6.11}$$

Eqs. (6.9) and (6.10) can be expressed in terms of the first and second differentials of $R(\tau)$ with time:

$$<\omega> = \frac{R'(0)}{jR(0)} \tag{6.12}$$

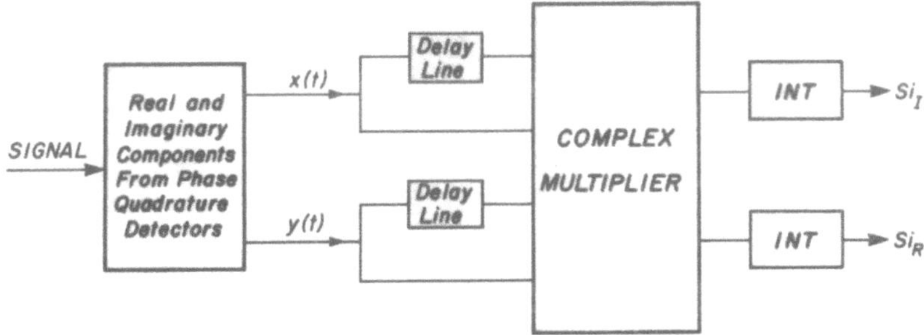

Figure 6.7. Block diagram of complex autocorrelator used for real-time Doppler color flow mapping. (Modified from Kasai et al. [7])

and

$$\sigma^2 = \left[\frac{R'(0)}{R(0)}\right]^2 - \frac{R''(0)}{R(0)}, \tag{6.13}$$

where primes $'$ and $''$ represent the first and second differential with respect to time. Because the computations by the above equations are time consuming, a further simplification was proposed by Kasai et al. [7]. For an exponential autocorrelation function

$$R(\tau) = |R(\tau)|e^{-j\phi(\tau)}, \tag{6.14}$$

with magnitude and phase as even and odd functions, respectively. Eqs. (6.12) and (6.13) can be transformed to

$$<\omega> = \frac{\phi(T)}{T} \tag{6.15}$$

and

$$\sigma^2 = \frac{2}{T^2}\left\{1 - \frac{|R(T)|}{R(0)}\right\}, \tag{6.16}$$

where T is pulse-repetition time. All the variables shown on the right side of the equations can be rapidly measured by a complex correlator for real-time imaging. The essentials of such a system are shown in the block diagram in Figure 6.7.

The real and the imaginary components of the quadrature-detected signal are divided into two, one of which is delayed in time by T. The original and the delayed signal are multiplied and integrated over time to yield Si_I and Si_R, which in turn are used to compute $< \omega >$ and σ^2 with Eqs. (6.15) and (6.16) and with the aid of the following formulas.

$$\phi(T) = tan^{-1}\left(\frac{Si_I}{Si_R}\right) \tag{6.17}$$

$$R(T) = \left[Si_I^2 + Si_R^2\right]^{\frac{1}{2}}, \tag{6.18}$$

$$R(0) = \int \left[X^2(t) + Y^2(t)\right]dt \tag{6.19}$$

The value of $< \omega >$ is used in the Doppler equation to determine the average value of blood velocity.

In essence, for every pixel of the interrogated region there are four computed parameters defining blood flow: direction of flow toward or away from the transducer, the mean flow velocity, and finally variance or turbulence in flow. The multiparametric information is stored in the image memory of the scan converter and is displayed in color superimposed on the B-mode gray scale image. The color scheme used to display parametric flow information can be understood with the aid of Figure 6.8.

The direction of flow is represented by the primary colors red and blue for the flow toward and away from the detector. The magnitude of flow (mean velocity) is represented by the brightness of the color. If eight shades of brightness for red and blue are used, then maximum velocity toward the detector will be represented by the top bin of the scheme shown in Figure 6.8. The variance in flow, σ^2, is displayed as green and is added to red and blue colors. Mixture of three colors is shown by the three overlapping circles. That is, increasing variance or turbulence in flow will be displayed by colors approaching cyan or yellow.

Color Doppler imaging is now regularly used in assessing various cardiac and vascular abnormalities and in the assessment of fetal circulation. The fact that these images concomitantly display functional and anatomic information has made this modality a valuable tool in clinical medicine [8].

6.3.4 Color Doppler for Induced Motions

The predominant use of ultrasound has been to monitor the natural motion of tissues. There is now a growing interest in stimulating tissue targets by controlled external mechanical vibrations and observing the response with ultrasound. Several schemes that use correlation between A lines and M-mode ultrasound have been proposed in the literature. Recently, Doppler imaging has been added to this list. Lerner et al. [9] proposed a method for characterizing stiffness of cancer tissues. The method is based on coupling audio band sound waves through a horn into the tissue and observing its motion with a commercial color Doppler instrument. Encouraging results have been reported when using

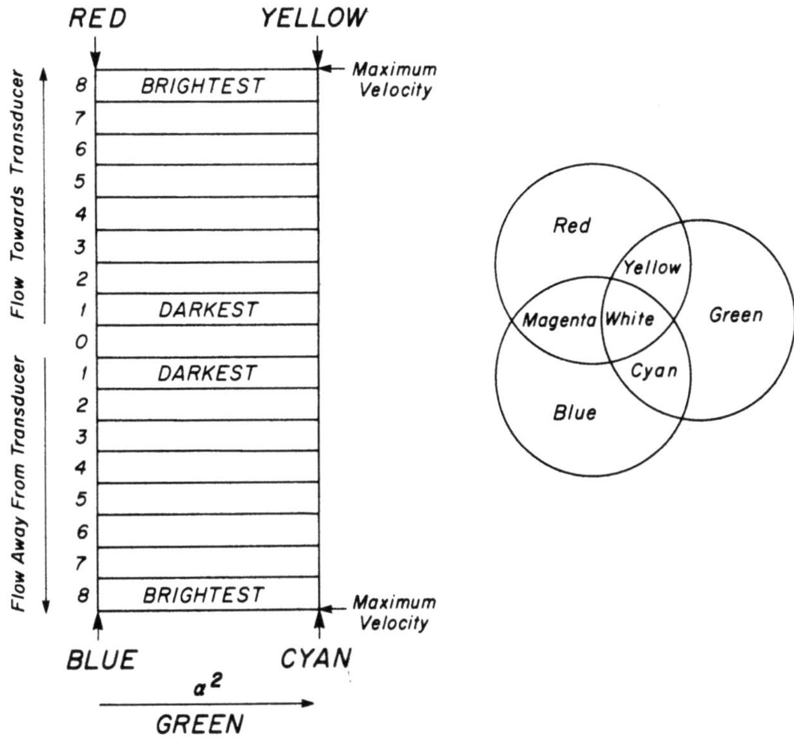

Figure 6.8. Graphic illustration of color scheme used to map flow information.

this method on phantoms and on tissues. A homogeneous phantom vibrated at suitable amplitude filled the color Doppler image. In contrast, a phantom containing hard regions simulating cancerous tissue, when subjected to such vibrations, appeared to be color free, thus providing differentiation from the soft tissue-like material.

6.4 Correlation Methods for Measuring Motion

Pulsed Doppler ultrasonic blood velocity imaging devices have problems with aliasing and with errors resulting from the lack of knowledge of the angle between the particle motion and the beam. New methods are being developed that do not have aliasing problems and measure the true velocity vector, including its direction. These methods use time domain correlation of echo signals or speckle correlation (also referred to as speckle tracking).

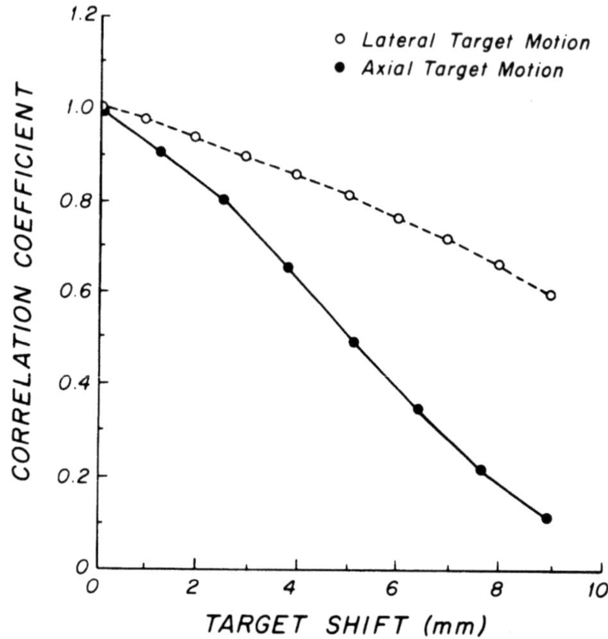

Figure 6.9. Images of speckle decorrelate over a length of about 2 mm axially and 5 mm laterally. (Reprinted with permission from The Institute of Electrical and Electronics Engineers. Copyright 1987 IEEE [10])

6.4.1 Speckle Tracking

The speckle of the image resulting from scatter from the blood cells is correlated in regions of the image between frames. This local correlation gives an estimate of the movement of the scatterers between frames if the movement was less than the decorrelation length. Trahey et al. [10] showed that the correlation coefficient of a speckle pattern from a nonmoving scattering phantom decorrelates significantly after about 2 mm in axial shift and 5 mm in lateral shift (Fig. 6.9).

By tracking the speckle, one can determine the motion of the particles [10], as long as the interframe time does not allow the scattering to decorrelate. Thus, one could track speeds up to 60 mm/s with a 30/s frame rate.

An example of a flow field calculated from two frames taken 1/30 of a second apart is shown in Figure 6.10.

6.4.2 Time Domain Correlation

The concept of time domain correlation technique for measuring flow velocity is simple and does not involve the use of Doppler phenomena. This technique has been proposed independently by several researchers [11, 12]. Two pulses

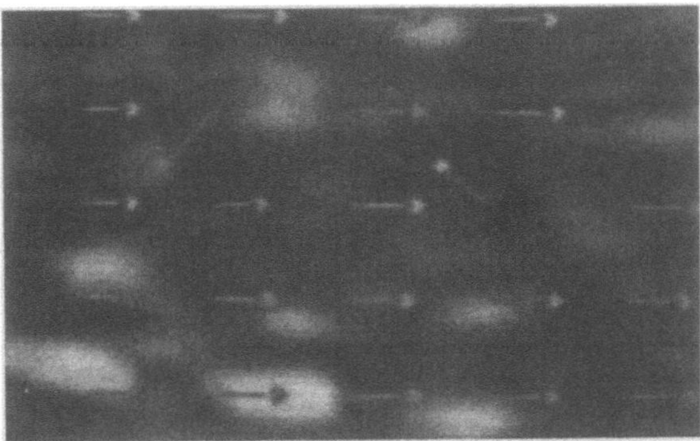

Figure 6.10. Flow field calculated from correlation of two video fields separated by 1/30 sec. (Reprinted with permission from The Institute of Electrical and Electronics Engineers. Copyright 1987 IEEE [10])

separated by time T are transmitted. The echoes of both the pulses are digitized and matched by scanning one with respect to the other. The time shift required so that the two echoes are maximally correlated indicates the time delay due to the movement ΔL of scatterers along the beam direction. If v is the velocity of particles,

$$\Delta L = vT. \tag{6.20}$$

If θ is the angle between the direction of the sound beam and the direction of flow,

$$\Delta L = \frac{\Delta d}{2cos\theta}, \tag{6.21}$$

where Δd is the extra distance the second sound pulse travels to meet scatterers that have moved by ΔL. If τ is the correlation time (equivalent to the transit time of the scatterer in the present case), the last two equations can be combined in the form

$$v = \frac{\tau c}{2Tcos\theta}. \tag{6.22}$$

To determine flow velocity, τ is measured by gating the reflected signal with a rectangular window. The range cell is shifted and overlaid onto the second signal repeatedly to determine the point of maximum correlation. Experiments based on this technique suggest that this method can be used to measure blood flow in vitro as well as in vivo [13]. However, the method suffers from limitations caused by refraction of ultrasound. Also, the interpulse time should not be so large as to allow scatterers to decorrelate. Turbulence in flow may also force the scattering to decorrelate faster than usual and may further complicate the measurements.

Table 6.1 Summary of Doppler methods.

Doppler ultrasonography	Advantages	Disadvantages
Continuous wave	High velocities can be measured	Uses two transducers Has no range resolution
Pulsed	Uses single transducer Measures blood flow at specific depths	Cannot measure very high velocities due to aliasing

6.5 Summary

Ultrasound is used in several modes to measure or to image motion (Class 4 scattering) of tissues and organs under normal and abnormal physiologic conditions.

M-mode ultrasound interrogates tissues by measuring the amplitude of an echo as pulses of ultrasound are repeatedly transmitted along a scan line.

Many Doppler methods are in clinical use. Each method has a specific place in the examination of normal and abnormal hearts and blood flow through vessels. The major advantages of continuous-wave and pulsed-Doppler modes are summarized below and in Table 6.1

Color Doppler uses multigate Doppler to produce a color map of flow and turbulence within a specified region. Multiple pulses are targeted along each scan line. The echoes are rapidly analyzed with an autocorrelator and Doppler information is displayed in real time as images. This method can be termed "functional imaging" because it shows the spatial distribution of flow information. Such a display is generally easy to comprehend. However, complex lesions may produce a multitude of flows in a small region. This appears as a complex color image that can be difficult to interpret. Construction of color images takes a substantial share of scanner time and decreases frame rates significantly. This can introduce temporal ambiguity in flow estimations. Color Doppler often suffers from aliasing artifacts. New methods based on signal correlation rather than Doppler phenomena are being developed to overcome aliasing problems. Doppler methods have proved to be valuable tools in the hands of practicing clinicians.

Bibliography

[1] S. B. Arvan, *Echocardiography: An Integrated Approach.* Churchill Livingstone. New York, NY, 1984.

[2] R. S. Adler, J. M. Rubin, H. Bland, and P. L. Carson, "Quantitative tissue motion analysis of digitized M-mode images: Gestational differences of

fetal lung," *Ultrasound in Medicine and Biology*, vol. 16, pp. 561–569, 1990.

[3] L. Hatle and B. Angelsen, *Doppler Ultrasound in Cardiology: Physical Principles and Clinical Applications*. Lea and Febiger. Philadelphia, PA, 1982.

[4] P. Atkinson and J. P. Woodcock, *Doppler Ultrasound and Its Use in Clinical Measurement*. Academic Press. New York, 1982.

[5] S. J. Goldberg, H. D. Allen, G. R. Mark, and R. L. Donnerstein, *Doppler Echocardiography*. Second ed. Lea and Febiger. Philadelphia, PA, 1988.

[6] K. Namekawa, C. Kasai, M. Tsukamoto, and A. Koyano, "Imaging of blood flow using auto-correlation," *Ultrasound in Medicine and Biology*, vol. 8, p. 138, 1982 (abstract).

[7] C. Kasai, K. Namekawa, A. Koyano, and R. Omoto, "Real-time two-dimensinal blood flow imaging using an autocorrelation technique," *IEEE Transactions on Sonics and Ultrasonics*, vol. SU-32, pp. 458–464, May, 1985.

[8] N. C. Nanda, *Textbook of Color Doppler Echocardiography*. Lea and Febiger. Philadelphia, PA, 1989.

[9] R. M. Lerner, S. R. Huang, and K. J. Parker, "Sonoelasticity" images derived from ultrasound signals in mechanically vibrated tissues," *Ultrasound in Medicine and Biology*, vol. 16, pp. 231–239, 1990.

[10] G. E. Trahey, J. W. Allison, S. M. Hubbard, and O. T. von Ramm, "Measurement of local speckle pattern displacement to track blood flow in two dimensions," *Proceedings IEEE Ultrasonic Symposium*, vol. 2, pp. 957–961, 1987.

[11] D. Dotti, E. Gatti, V. Svelto, A. Ugge, and P. Vidali, "Blood flow measurements by ultrasound correlation techniques," *Energia Nucleare*, vol. 23, pp. 571–575, November, 1976.

[12] S. G. Foster, P. M. Embree, and W. O'Brien, Jr., "Flow velocity profile via time-domain correlation: error analysis and computer simulation," *IEEE Transactons on Ultrasonics Ferroelectrics and Frequency Control*, vol. 37, pp. 164–175, May, 1990.

[13] P. Embree and W. O'Brien, Jr., "Volumetric blood flow via time-domain correlation: experimental verification," *IEEE Transactions on Ultrasonics Ferroelectrics, and Frequency Control*, vol. 37, pp. 176–189, May, 1990.

7
Backscatter-Imaging Instruments

7.1 Introduction

Instruments can be divided into several groups depending on the type of transmit insonation and signal reception used. The principal mode of imaging in medicine is echo or backscatter geometry. This has evolved because of the limited number of available windows into the body. Ultrasonic energy is stopped by bones and gas-filled cavities which can be found throughout the body. Therefore, medical instruments have evolved that transmit energy into and receive backscattered signals out of the same window. Transmission geometries in which the energy is transmitted entirely through the body, similar to x-ray techniques, have not been successful yet, although several have been developed and will be described in the next chapter. Other types of geometry such as orthogonal scattering, mode conversion scattering, and nonlinear imaging have not been successfully applied and will not be discussed in this book. This chapter begins with a brief review of piezoelectrics, beamforming, and scanning and then describes the relationship of scattering hierarchies to images obtained in the echo mode.

7.2 Transducers

7.2.1 Piezoelectrics

The active components of transducers shown in Figure 7.1 are made from piezoelectric materials [1], which are dielectric materials that produce charge when they are subjected to strain or which produce strain when the electric field distribution across them is altered. Thus, piezoelectric materials can be used to create force from an electric signal (transmit) or to produce an electric signal from an externally applied force (receive). The unpolarized piezoelectric ceramic (Fig. 7.2) consists of a distribution of domains that are oriented randomly. Each domain has an electric dipole. To polarize a piezoelectric transducer, it is usually heated so that the domains can rotate, and then an electric field is placed across the transducer. The domains rotate into an orientation that depends on the sign of the electric field distribution across the transducer. The transducer is then cooled to "freeze" the domains. The result is a polarized transducer. Because the domains are oriented, the transducer is now piezoelectric. When the material is subjected to force, the domains rotate, changing the charge distribution between the electrodes. On the other hand, if an electric field is placed across the transducer, the domains tend to align with the electric field, changing the size of the transducer.

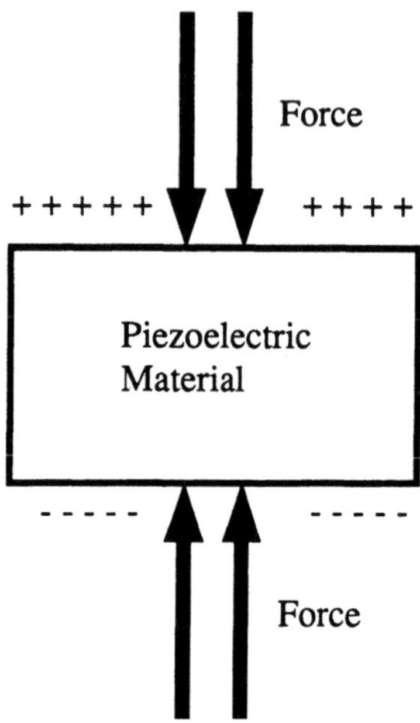

Figure 7.1. Piezoelectric transducers produce charge when strained or they produce strain when submitted to an electric field.

7.2.2 Lenses

There are two general ways to focus ultrasound using individual piezoelectric elements. One is to place a lens on the front surface of the transducer and the other is to cut the element or to machine the element into the appropriate shape (Fig. 7.3).

7.2.3 Construction

Figure 7.4 illustrates the piezoelectric element and its placement inside a single-element transducer. This is a cross section through the transducer and includes matching layers placed on the front of the piezoelectric element [2]. The matching layers are used to transform the very high acoustic impedance of the piezoelectric ceramic to that of biologic tissues. This results in the same effect as matching layers on the surface of lenses in expensive optical cameras, which are designed to inhibit reflection within the lens system caused by high impedance changes between air and the glass lens. The piezoelectric ceramic is sometimes backed with an absorbing material, so that energy coming into the piezoelectric element does not reflect back and forth within the element. Instead, the energy goes straight through the element into the backing material without reflecting. In general, the matching layers are a series of quarter-wave matching layers that are tuned to cause the least change in acoustic impedance as the

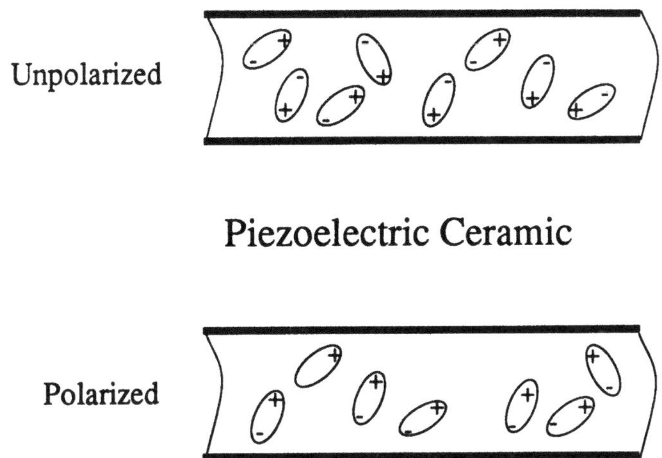

Figure 7.2. Domains in unpolarized ceramic are unorganized. Polarization organizes the domains, making the material piezoelectric.

wave enters the transducer. If the matching layers are especially good and result in very little reflection at the front surface of the transducer, air backing can be used, thereby eliminating a 3–dB loss of energy into the backing material. Modern transducers are generally air backed and very well matched from the front of the transducer, so that the ultrasonic energy does not vibrate back and forth within the transducer [3].

7.3 Beams

The purpose of the transducer is to produce a narrow beam of insonation which acts as a flashlight that insonifies small cylindrical regions of the tissue. Because the transducer is required to operate in a pulse-echo mode, the resolution of the imaging system can be divided into two parts: the lateral resolution and the axial resolution.

7.3.1 Lateral Resolution

The lateral resolution of an ultrasound transducer depends on the size of the aperture in wavelengths. The aperture is the physical extent of the transducer perpendicular to the beam. The larger the transducer, the better the focus in general. The shape of the aperture defines the shape of the beam—e.g., a circular aperture produces a symmetrical beam. A linear or rectangular aperture produces a beam with better resolution in the plane of the largest aperture extent

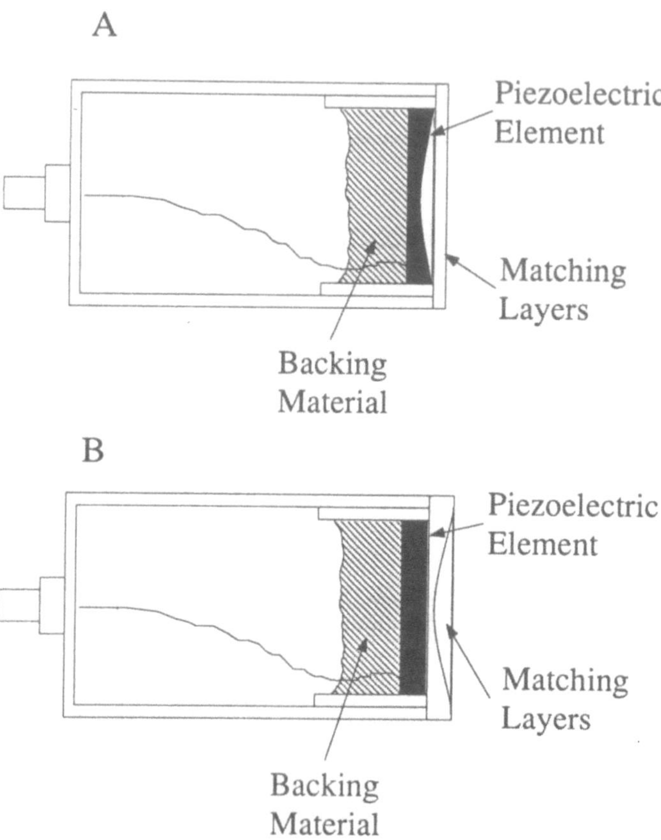

Figure 7.3. Schematic of a cross section through a transducer. In A, a curved transducer is used to focus and in B a lens is used.

and less resolution out of that plane because of the differences in size of the aperture. Figure 7.5 illustrates a beam produced by a transducer. The beam has a focal region whose width depends on the size of the aperture in wavelengths and the numerical aperture L/a. The width (full width half maximum) of the focal region is described by the following equation

$$FWHM_L = \frac{\lambda L}{a}, \tag{7.1}$$

where λ is the wavelength of the center frequency, L is the focal length, and a is the diameter of the transducer [4].

7.3.2 Axial Resolution

resolution depends on the bandwidth of the transducer but not on the center frequency. The equation for axial resolution in mm is

$$FWHM_A = 1.37/\Delta f, \tag{7.2}$$

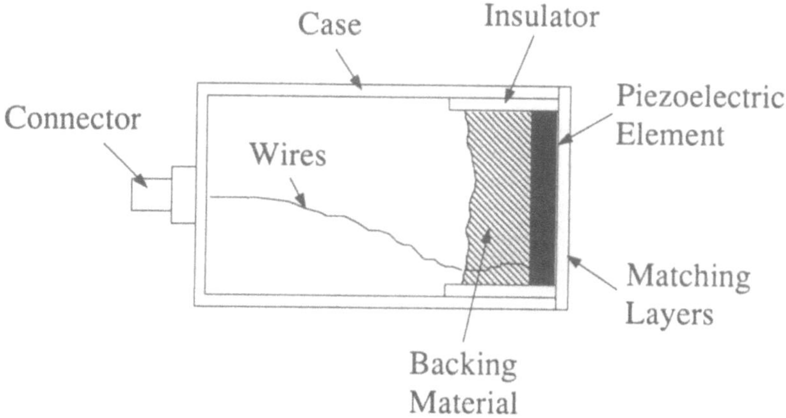

Figure 7.4. Cross section through a transducer shows the piezoelectric element and associated parts.

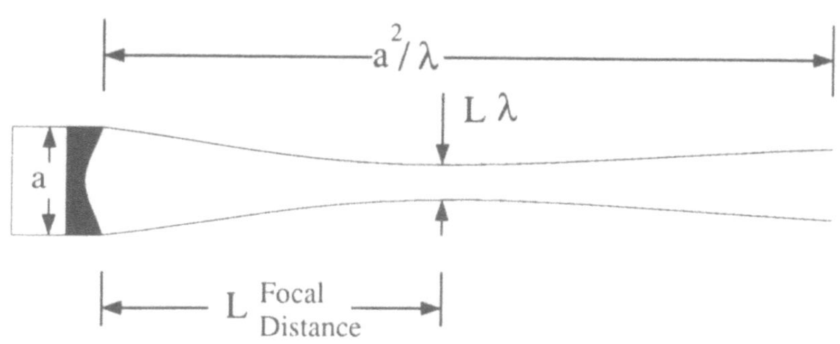

Figure 7.5. Schematic of focusing with a transducr. The lateral width of the beam depends on the aperture a, the focal length L, and the wavelength λ.

where Δf is the bandwidth of the system in MHz.

The broader the bandwidth, the better the axial resolution of the imaging system independent of the center frequency. This requires that the piezoelectric element and its associated mounting components in the transducer be designed to inhibit ringing. Typical bandwidths for modern transducers are greater than 50% of their center frequency.

7.3.3 Resolution Cell

The resolution cell is the size of the impulse response of the imaging system. It can be considered to be the product of the axial, lateral, and transverse resolutions. The resolution cell defines the smallest bright area that will occur in the image. The image is the magnitude of the sum of wavelets representing the resolution cells.

7.4 Mechanical Scanning

There are two ways to scan a transducer over a region of interest to obtain a planar image or tomogram in the echo mode. One is to use mechanical motion and the other is to electronically alter the phasing of the acoustic pulses across the aperture to guide the beam in various directions. In general, circular apertures are mechanically scanned or wobbled back and forth so their beams traverse a wedge-shaped region in a plane. This results in a wedge-shaped image. The advantage of such a method is that very little electronic phasing circuitry is required. The disadvantage is that the frame rate is dependent on the speed with which the transducer can be wobbled. Mechanical scanners consist of a single fixed-focus transducer that is physically moved across an aperture or rotated back and forth about a central axis. The system transmits energy periodically, receiving backscatter from a narrow region representing the effective beam for each transmitted pulse. For each transmit point, the received signal is processed and demodulated, and its envelope as a function of time is used to produce a line in the image that has brightness proportional to the magnitude of the backscatter as a function of depth along the beam. A sequence of adjacent lines in the image results from repeating this process as the transducer is moved.

7.5 Lensless Beamforming

7.5.1 Phased Array

The phased-array scanner consists of an array of elements covering a linear aperture with or without curvature. In transmit focusing, the elements are phased to produce a focused beam of insonifying energy (Fig. 7.6). In receive focusing, the phase of the signal from each element of the array is shifted through the object to focus near the region from which the echoes are arriving (Fig. 7.7). Therefore, the focal point of the receive aperture is moved at one-half the speed of sound into the tissue in concurrence with reflected energy coming back from the transmitted pulse. Some instruments require a large window because they use a long linear array, thus limiting the method to regions of the body having large windows such as the abdomen. Other instruments use a small aperture and sector scanning to take advantage of small acoustic windows such as the space between the ribs. The advantage of these scanners over mechanical scanners is

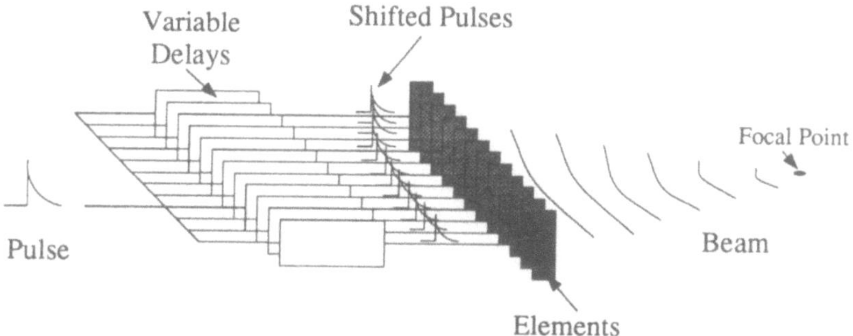

Figure 7.6. Energy exiting from the array is phased so as to focus at a specified point.

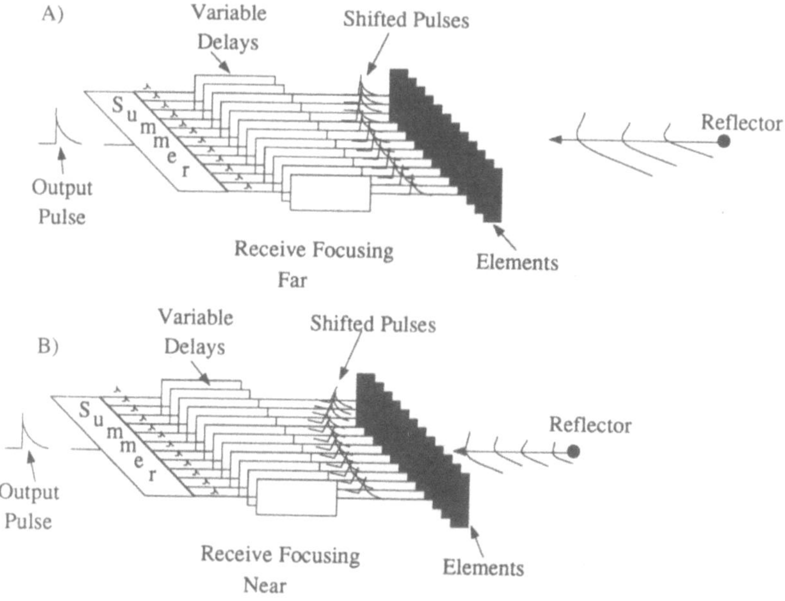

Figure 7.7. (A) Electronic focusing for a point far from the array. The delays are set up to mimic the shifts in a lens focused at the far point. (B) The delays are set up to focus on a point near the array.

that they can produce images at a high frame rate and there are no mechanical parts. The disadvantage is that the out-of-plane focus is not usually as good as with circular apertures in mechanical scanners.

The term "phased array" is somewhat a misnomer but it has come to mean an array that produces a wedge-shaped image by phasing the elements. This is a one-dimensional array which produces a set of beams in a fan primarily so the imaging can be done through narrow windows such as the space between the ribs.

7.5.2 Annular Array

Very often focusing is achieved through the use of an acoustic lens or concave spherically shaped radiators, methods that could be termed "geometric focusing." The concept of phased annular arrays in medical ultrasound was probably first proposed by Reid and Wild in 1957 [5]. In general, a planar piezoelectric disk is divided into a number of concentric annular elements. Each element is provided with an independent, variable time delay, permitting the transmit and receive foci to be moved along the axis of the transducer. This results in a symmetric, nearly diffraction-limited focus at all imaging depths. Many annular array systems have been made over the years [6–8]. A lens on the front of an annular array greatly reduces the range of phase shifts required across the array to maintain focusing. The advantage of the annular array is that the impulse response is symmetric and that the out-of-plane resolution is as good as in-plane resolution. The disadvantage of phased annular arrays is that they must be mechanically scanned.

7.5.3 Diffractionless Array

Conventionally, transducers with Gaussian amplitude shading and Fresnel phase shading are used in medical imaging because the beam patterns are smoother in both near and far fields [1]. The Gaussian beam diverges dramatically after the Rayleigh distance (focal region) because of diffraction. Dynamic focusing is used to improve lateral resolution of the Gaussian beam transducer in pulse-echo imaging. However, these transducers can be focused only at one point in transmit mode. The transmit beams diverge from the focal point, thus reducing resolution. The use of a transmit beam that does not diverge throughout the region of interest would greatly improve current medical imaging methods.

Recently, Durnin [9] discovered a nondiffracting beam that can have long depth of field. Like the Gaussian beam, the nondiffracting beam has a small center lobe but relatively large side lobes which travel in parallel with the center lobe. Since the nondiffracting beam was discovered, experiments for physical realization of finite aperture J_0 beams have been done [10].

Lu and Greenleaf [10] have described the use of these beams for medical imaging with an annular array. Deep depth of field was obtained with a narrow beam but relatively high side lobes. Because of the long depth of field (>200 mm), the J_0 Bessel nondiffracting transducer could be used to replace the multiple transmissions required in conventional mechanical B-scanners, which reduce the imaging frame rate and result in blurred images of moving objects. The small size of the center lobe of the J_0 Bessel nondiffracting transducers could also be used in transmission imaging, such as transmission ultrasonic tomography, with the expectation of obtaining higher imaging quality over the conventional Gaussian beam transducer which must be focused at a fixed distance. Another potential application of the J_0 Bessel nondiffracting transducer would be tissue characterization, which requires correction for diffraction in the estimation of tissue attenuation or B/A (nonlinear parameter) [11–13].

7.6 Effect of Scattering Class on Image

7.6.1 Class 0 Absorption and Speed

The effect of absorption and speed on ultrasonic images is a bulk effect producing variations in brightness and variations in position of reflectors in the image. As shown in Figure 7.8, attenuation within a lipoma (mass consisting of fat) is about equal to that of other regions of the image, resulting in neither a brightened region behind the lipoma nor a darkened region [14]. However, the appearance of the diaphragmatic border is shifted by the lower speed of sound through the lipoma, another Class 0 effect produced by variations in acoustic speed. Background speed of sound is generally assumed by commercial instrument makers to be on the order of 1,560 m/s and background attenuation is generally assumed to be on the order of 0.5 dB/cm per MHz. Modern instruments do not allow for variations of the assumed speed of sound; however, they do attempt to correct for attenuation through the use of time-variable gain. Time-varying gain is produced with programmable amplifiers that increase gain exponentially with time so that attenuation of the echoes returning from deeper into the tissue is corrected.

7.6.2 Class 1 Speckle

Figure 7.8 illustrates the effect of speckle or Class 1 scattering on acoustic images. One can see a uniform distribution of gray spots in the background of this image of the liver, which is the effect of speckle on the image. The number of scatterers per resolution cell is high and, therefore, speckle is fully developed within the lipoma [4].

Speckle results from interference of many statistically independent phasors whose amplitudes and phases are independent. If the real and imaginary parts of the resulting pattern are independent and follow a zero mean Gaussian probability distribution with equal variances, then the resulting envelope will have a Rayleigh statistical distribution, as described in Chapter 5. The ratio of the mean to the variance of this envelope will be around 1.91. In fully developed speckle, the characteristics of the tissue only affect the mean echo level. If the number density of scattering sites per resolution cell is low, then the image characteristics will be dependent on the density, and associated tissue characteristics can be estimated with statistical methods [15].

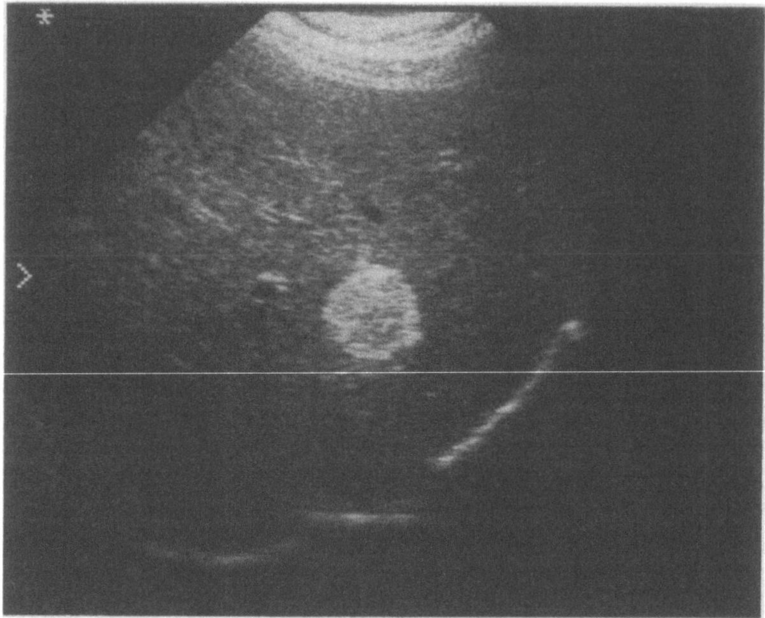

Figure 7.8. Transverse oblique sector scan of liver with a lipoma (fat tumor). Variation in speed (Class 0) in the tumor is indicated by the shift in position of the specular reflection (Class 3) of the diaphragm below the tumor. Increased speckle scatter (Class 1) is shown within the tumor. Various vessels and other resolved reflectors (Class 2) appear elsewhere within the liver. (Reproduced with permission from Radiological Society of America [14])

7.6.3 Class 2 Resolved Scatterers

The Rayleigh distribution of the received envelope will be altered if there are resolved scatterers present in the imaged field. If the speckle scatterers (Class 1) are accompanied by resolved scatterers (Class 2), then the statistical distribution of the envelope will be altered and the ratio of mean to variance will be increased (Chapter 5, section 5.2.3). If structural order such as periodicity is present in the tissue, then autocorrelation analysis can be used to evaluate the associated features of the tissue (see Chapter 5). Figure 7.8 includes the effect of resolved scatterers within the liver where both speckle and resolved scatterers appear.

7.6.4 Class 3 Specular Scatterers

Specular scattering results in large reflections that greatly alter the autocorrelation function of the scatter patterns. Because the specular reflector produces clear pulse reflections, they can be detected using signal analysis [16]. Figure 7.8 includes reflections from the surface of the diaphragm which are very clear and result from the target extending through the width of the beam and producing

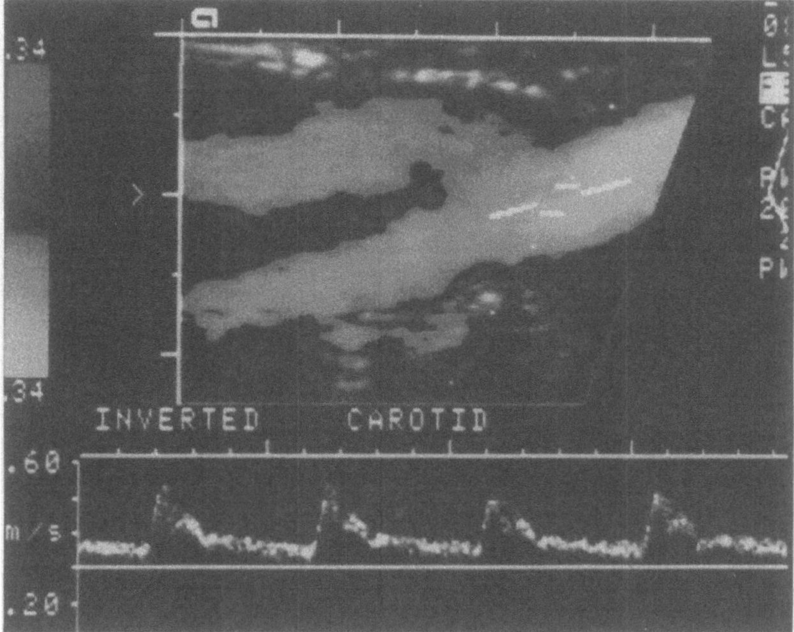

Figure 7.9. Longitudinal color Doppler sonogram of the carotid artery. Flow velocity is coded in color. A sample volume, indicated by the white lines, is interrogated by continuous-wave Doppler and yields flow velocity as a function of time (bottom panel).

a signal that has very little interference, much like that from a mirror. If the surface of the organ is perfectly smooth, to within one tenth of a wavelength or so, then the reflections will be purely specular and most of the surface will not be seen because the energy will probably be reflected away from the receiver.

7.6.5 Class 4 Motion

Motion is perceived with modern imaging instruments in two ways: through real-time high frame rate cine depictions of the object and through the Doppler shift of the backscattered signals. The Doppler shift is detected by two methods: continuous-wave and pulsed Doppler. Imaging of motion can be done using a method called "color Doppler" in which Doppler shift is mapped onto a color scale and superimposed on the B-scan (Fig. 7.9). The advantage of this method is that motion of fluids or solid tissues can be estimated quantitatively in real time. These methods have been used to evaluate cardiovascular abnormalities [17, 18], tumor blood flow [19], and tissue stiffness [20, 21].

7.7 Conclusion

Modern clinical ultrasound imaging instruments, working in the backscatter mode, image all five classes of scattering. The range of frequencies normally used for medical macroimaging, as opposed to microimaging, results in the scattering classes being related to biologic classes, as described in Chapter 1. Ultrasound imaging instruments provide real-time tomography with good resolution and with good contrast between various soft tissues. Continued research on the characteristics of scattering signals and their relationship to tissue state should provide methods of separating the various biologic classes.

Bibliography

[1] G. S. Kino, "Wave propagation with finite exciting sources," in *Acoustic Waves: Devices, Imaging, and Analog Signal Processing*, pp. 154–317, Prentice-Hall, Inc. Englewood Cliffs, NJ, 1987.

[2] G. Kossoff, "The effects of backing and matching on the performance of piezoelectric ceramic transducers," *IEEE Transactions on Sonics and Ultrasonics*, vol. 13, pp. 20–30, 1979.

[3] F. S. Foster and J. W. Hunt, "The design and characterization of short pulse ultrasound transducers," *Ultrasonics*, vol. 16, pp. 116–122, 1978.

[4] R. F. Wagner, S. W. Smith, J. M. Sandrik, and H. Lopez, "Statistics of speckle in ultrasound B-scans," *IEEE Transactions on Sonics and Ultrasonics*, vol. 30, pp. 156–173, 1983.

[5] J. M. Reid and J. J. Wild, "Current developments in ultrasonic equipment for medical diagnosis," *IRE Transactions on Ultrasonic Engineering*, vol. 5, pp. 44–56, 1957.

[6] H. E. Melton, Jr. and F. L. Thurstone, "Annular array design and logarithmic processing for ultrasonic imaging," *Ultrasound in Medicine and Biology*, vol. 4, pp. 1–12, 1978.

[7] F. S. Foster, M. Arditi, M. S. Patterson, D. Lee-Chahal, and J. W. Hunt, "Breast imaging with a conical transducer/annular array hybrid scanner," *Ultrasound in Medicine and Biology*, vol. 9, pp. 151–164, 1983.

[8] D. R. Dietz, S. I. Parks, and M. Linzer, "Expanding-aperture annular array," *Ultrasonic Imaging*, vol. 1, pp. 56–75, 1979.

[9] J. Durnin, "Exact solutions for nondiffracting beams. I. the scalar theory," *Journal of the Optical Society of America (A)*, vol. 4, pp. 651–654, 1987.

[10] J. y. Lu and J. F. Greenleaf, "Pulse-echo imaging using a nondiffracting beam transducer," *Ultrasound in Medicine and Biology*, vol. 17, pp. 265–281, 1991.

[11] M. Insana, J. Zagzebski, and E. Madsen, "Improvements in the spectral difference method for measuring ultrasonic attenuation," *Ultrasonic Imaging*, vol. 5, no. 4, pp. 331–345, 1983.

[12] M. A. Fink and J. Cardoso, "Diffraction effects in pulse-echo measurement," *IEEE Transactions on Sonics and Ultrasonics*, vol. SU-31, pp. 313–329, July, 1984.

[13] T. Sato, A. Fukusima, N. Ichida, H. Ishikawa, H. Miwa, Y. Igarashi, T. Shimura, and K. Murakami, "Nonlinear parameter tomography system using counterpropagating probe and pump wave," *Ultrasonic Imaging*, vol. 7, pp. 49–59, 1985.

[14] C. C. Reading and J. W. Charboneau, "Case of the day. Ultrasound. Hepatic lipoma," *Radiographics*, vol. 10, pp. 511–512, 1990.

[15] J. M. Thijssen, "Ultrasonic tissue characterization and echographic imaging," *Medical Progress Through Technology*, vol. 13, pp. 29–46, 1987.

[16] N. M. Bilgutay, K. D. Donohue, and X. Li, "Nonparametric flaw detection in large gained materials," *Proceedings IEEE Ultrasonics Symposium*, vol. 2, pp. 1137–1141, 1990.

[17] N. Wittlich, R. Erbel, M. Drexler, S. Mohr-Kahaly, R. Brennecke, and J. Meyer, "Color-Doppler flow mapping of the heart in normal subjects," *Echocardiography*, vol. 5, pp. 157–172, 1988.

[18] J. K. Oh, B. K. Khandheria, J. B. Seward, W. K. Freeman, L. J. Sinak, and A. J. Tajik, "Transesophageal color flow imaging," *Echocardiography*, vol. 5, pp. 407–416, 1988.

[19] P. N. T. Wells, M. Halliwell, R. Skidmore, A. J. Webb, and J. P. Woodcock, "Tumour detection by ultrasonic Doppler blood-flow signals," *Ultrasonics*, vol. 15, pp. 231–232, 1977.

[20] K. J. Parker, S. R. Huang, R. A. Musulin, and R. M. Lerner, "Tissue response to mechanical vibrations for sonoelasticity imaging," *Ultrasound in Medicine and Biology*, vol. 16, pp. 241–246, 1990.

[21] M. Tristam, D. C. Barbosa, D. O. Cosgrove, J. C. Bamber, and C. R. Hill, "Application of Fourier analysis to clinical study of patterns of tissue movement," *Ultrasound in Medicine and Biology*, vol. 14, pp. 695–707, 1988.

8
Computed Transverse Imaging

8.1 Computed Tomography

The first computed ultrasonographic tomograms (CT) were produced in 1973 [1]. Since then, considerable ingenuity has been demonstrated by many researchers in implementing a wide variety of CT imaging techniques. These techniques can be divided into two broad categories. One category of methods is based on the straight-ray approximation for either transmission or reflection measurements. Because the straight-ray approximation is used, the measurements are regarded as projections or shadows. The second category of methods results from attempts to take diffraction into account by mathematically inverting an approximate wave equation. It is known as diffraction tomography and is discussed later.

Adaptive processing [2] and speckle processing [3] represent *a posteriori* approaches to imaging an inhomogeneous medium. Although potentially useful, they as well as incoherent imaging [4] are mentioned here but not discussed in detail because they are too esoteric for this book.

8.1.1 Reconstruction From Transmission Projections

The straight-ray approximation was used in early attempts at CT, the equations for which will be briefly discussed here. Beginning with Eq. (3.1) with $\gamma_\rho = 0$, we have

$$\nabla^2 P(\hat{r},\omega) + K^2(1 + \gamma_\kappa)P(\hat{r},\omega) = 0, \tag{8.1}$$

where the complex value $1 + \gamma_\kappa$ is commonly referred to as the complex refractive index. If the signal P is harmonic, i.e.,

$$P(\hat{r},\omega) = \psi_o \exp\left[jK\phi(\hat{r})\right], \tag{8.2}$$

we obtain from Eq. (8.1)

$$iK^{-1}\nabla^2\phi - \nabla\phi \cdot \nabla\phi + (1 + \gamma_\kappa)^2 = 0, \tag{8.3}$$

where $\phi(\hat{r})$ is the acoustic path length or eikonal. If $K \to \infty$, then we have

$$\nabla\phi \cdot \nabla\phi = (1 + \gamma_\kappa)^2. \tag{8.4}$$

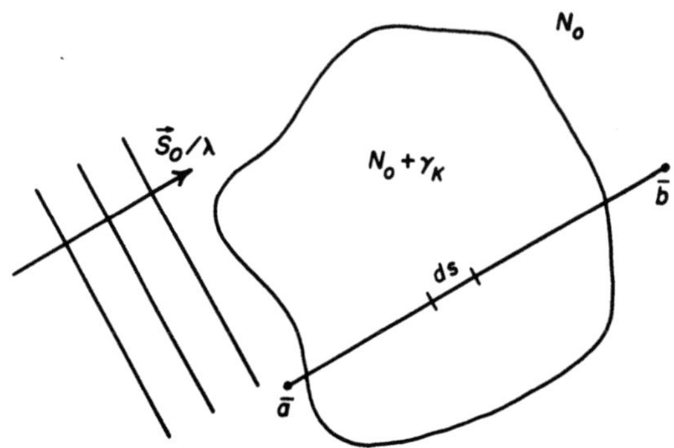

Figure 8.1. Geometry for transmission tomography. A continuous wave with vector $\vec{S}_o/\lambda$ travels in a medium with constant refractive index N_0 and enters an object with refractive index $N_0 + \gamma_\kappa$. Rays have initial direction $\vec{S}_o$, incremental length ds, and may or may not be considered straight, depending on the propagation model being considered. (Modified from [7])

If $\phi = \phi_o + \phi_1$, where ϕ_0 is the solution in the unperturbed medium ($\gamma_\kappa = 0$), we have $\nabla\phi_o \cdot \nabla\phi_o = 1$. If $(\gamma_\kappa)^2$ and $\nabla\phi_1 \cdot \nabla\phi_1$ are negligible, we have from Eq. (8.4)

$$\nabla\phi_o \cdot \nabla\phi_1 = \gamma_\kappa. \tag{8.5}$$

The unperturbed plane wave $\phi_o = \vec{S}_o \cdot \hat{r}$, giving from Eq. (8.5)

$$\vec{S}_o \cdot \nabla\phi_1 = \gamma_\kappa. \tag{8.6}$$

This can be integrated to yield

$$\phi_1\left(\hat{a}\right) - \phi_1\left(\hat{b}\right) = \int_{\hat{a}}^{\hat{b}} \left(\gamma_\kappa\right) ds, \tag{8.7}$$

where ds is along the path parallel to $\vec{S}_o$.

ϕ_1 is complex: the imaginary part relates to attenuation and the real part to speed.

Obtaining these data $\left[\phi_1\left(\hat{a}\right) \text{ and } \phi_1\left(\hat{b}\right)\right]$ involves transmission measurements to obtain projections in a fashion similar to that of x-ray CT [1, 5, 6]. Transmitting and receiving transducers are typically located on opposite sides of the subject under investigation, as indicated in Figure 8.1. The measurements are confined within the plane of interest either by the geometric acoustic response of the transducer or by subsequent signal-processing operations. Each translation of the transmitter/receiver pair produces a sequence of measurements called a projection. Projections are obtained at many angles of view by rotating either the object or the line of translation through small angular steps and repeating the translation or scan at each angle. This method was also used in so-called "first-generation" x-ray CT scanners. The method is time-consuming and has been abandoned in the x-ray field. Data collection can be speeded with transducer arrays or so-called "fan beam" (second generation or later) geometries, or both [8, 9].

After a full set of projections is obtained, an image can be reconstructed by any of various CT reconstruction methods [10]. Reconstruction is a term that refers to the solution of Eq. (8.7) for the desired physical property γ_κ or γ_ρ. The reconstruction process may be regarded as the problem of solving a set of linear simultaneous equations that could, in principle, be solved by matrix inversion [11]. However, because matrices with ranks of thousands are involved [12], only computationally efficient algorithms can be used practically. One of the earliest algorithms was termed "algebraic reconstruction technique" (ART) [1, 13–15]. The ART approach is essentially iterative and involves modifying successive estimates of the image until it is consistent with the measured projections. ART has the advantage of easily incorporating any *a priori* information and is relatively tolerant of noisy, missing, or erroneous projections. However, it is not easy to know when to terminate ART because the estimates can get worse after a certain number of iterations [10]. More recently, however, the convolution/back-projection approach [10, 16, 17] is used almost exclusively [8]. The choice of which method to use is usually determined by convenience factors in straight-line CT.

Speed Measurements.   Projections relating to several different ultrasonic variables can be measured and reconstructed into images. It is well known that phase shift or transmission time-delay measurements can be made with high accuracy [8, 18, 19]. Although the analysis presented in Section 8.1 was for a single frequency, in practice, wide-bandwidth transmissions are used to estimate time delay and, hence, acoustic sound speed. The required assumption implicit in using wide-bandwidth transmissions is that tissues exhibit negligible velocity dispersion (i.e., speed dependence with frequency). Experience shows that this assumption is reasonable [20]. Narrow pulses are often used because these allow time delays to be estimated by simple signal processing. For example, after the pulse is transmitted, a clock is started. The arrival of the pulse is determined by a threshold detector in the receiver. The threshold is set just above the noise level and once exceeded (by the arrival of the pulse), the clock is stopped, yielding the "time-of-flight" measurement.

Time-of-flight measurements require high precision, and clock rates are often of the order of 100 MHz. Because the bandwidth is always limited in practice (usually by transducer characteristics), the pulse always has a finite rise time and typically consists of several oscillations. Variations in attenuation along different ray paths cause fluctuations in received signal amplitude and, hence, errors in the estimation of time delays (due to "time walk" and "time hop" phenomena [21]) when simple threshold detection is used. Some of these difficulties can be avoided by calibrating the time-delay measurement with respect to received amplitude or by determining the time of the first zero crossing after the threshold is exceeded [6]. To allow lower peak-power levels or higher signal-to-noise ratios, or both, large time • bandwidth product signals and matched filter or pulse compression receivers can be used [5, 18, 22].

Attenuation. Projections relating to pure absorption cannot be measured easily. The attenuation of sound as it propagates through tissue is due to absorption and scattering (including reflection) as well as secondary effects such as mode conversion and harmonic generation caused by nonlinearities, as described in Chapters 4 and 5 [23]. For narrow-bandwidth measurements, it is not possible to separate the individual contributions to the overall attenuation of the signal. Therefore, only attenuation, rather than absorption, reconstructions are made. Both the peak amplitude and the energy of the received signal provide measures of attenuation. However, considerable care is required to obtain meaningful attenuation projections. Refraction can cause the beam to have other than normal incidence at the receiver, resulting in destructive interference known as "phase cancellation" [24] and producing overestimation of the actual attenuation. Phase-insensitive or nonlinear transducers, which respond only to intensity or to absolute value of the pressure, may alleviate this problem [24, 25]. Better sensitivity can be achieved with a receiving transducer made from a fine array of piezoelectric elements and use of incoherent signal processing [26]. The receiving transducer must be large enough so that the beam is never deflected beyond the limits of its active surface [27]. However, because of these caveats, attenuation projections obtained by straightforward calculations of the received energies of the pulse transmissions are fundamentally in error [28].

Attenuation Coefficient. The attenuation coefficient, which is analogous to the absorption coefficient in Eq. (8.7), is an approximately linear function of frequency in tissues [29]. However, the frequency dependence of attenuation due to scattering varies from being proportional to a square law for Class 2 scatterers to linear frequency independence for Class 3 reflectors, as described in Chapters 4 and 5 [30]. By considering only the slope of the attenuation frequency dependence, the constant component of attenuation due to specular (Class 3 scattering) reflection is removed from the attenuation projection. The resultant measure is termed the "integrated attenuation coefficient" [28]. The advantage of excluding reflection effects from the measurement is that reflection exhibits strong extrinsic anisotropy [31], resulting in inconsistent sets of projections and erroneous reconstructions. Several techniques have been proposed to estimate the integrated attenuation coefficient [28, 32]. Perhaps the simplest is the frequency-shift method [28] which, assuming the frequency dependence is linear, relates the integrated attenuation coefficient to the shift in signal center frequency after propagation through tissue. A further advantage of reconstructing the integrated attenuation coefficient is that the method is potentially insensitive to variations in the signal level caused by refraction.

Nonlinearity. Water-based tissues are as nonlinear (a Class 0 effect) as water; fat, in particular, is very nonlinear [33]. Images can be reconstructed of the so-called nonlinear parameter, B/A [34]. One manifestation of nonlinearity is a change in propagation velocity with pressure. In principle, projections relating to the nonlinear parameter could be obtained by measuring differences in time-of-flight projections caused by variations in static pressure. More sophisticated approaches involve measuring the interaction of high frequency probing waves with low frequency "pump" waves to obtain spatial Fourier-transform relationships [34].

8.1.2 Reconstruction From Reflections

Projections obtained from measurements of scattering in directions back toward the source of insonation are used in reflection tomography. Reflection tomography also can be based on the straight-ray approximation. However, the direction of integration for reflection tomography is orthogonal to the propagation direction. Consider a small omnidirectional transducer transmitting a short-duration pulse into a medium in which the acoustic velocity is nearly constant. Echoes returning to the transducer at a particular instant in time could originate from scatterers located anywhere on a spherical surface. The amplitudes of the echoes depend on the backscatter strengths or reflectivity and the attenuation along the various propagation paths. Therefore, such a system provides a means of estimating the integrals of reflectivity over concentric spherical surfaces [35, 36]. The radius of each sphere is proportional to the echo delay and the returning train of echoes may be regarded as a type of projection.

If separate transmit and receive transducers, instead of a single transducer, are used, then the surfaces of integration are elliptical. It has been shown [35, 37] that an image of reflectivity can be reconstructed from a set of either type of surface integral (obtained from many different views) in a fashion similar to the conventional straight-line convolution/backprojection method mentioned in the preceding section. Of course, the backprojection is performed along curved paths. Furthermore, if an appropriate "optimum" pulse shape is transmitted, then the convolution operation is not required. If a large plane wave transducer is used instead of one or two "point" transducers, then the integrals are over planes and straight-line reconstruction methods can be used [38]. Provided the straight-ray assumption holds, the resolution of these techniques is theoretically limited by the transmitted bandwidth and not by the transducer beam width [35, 36].

When backscatter measurements over only one surface or angle of view are available, such as in the acoustic probing of geologic formations, "seismic" methods can be used to reconstruct an image [39]. Seismic imaging essentially involves the backward propagation or "migration" of the measured field into the object. This is achieved by phase filtering the angular spectrum of the measured field [40] and requires the assumption of constant propagation velocity or of knowledge of propagation velocities. However, allowances for known velocity differences in a stratified medium are possible because the backward propagation can proceed layer by layer.

A variation of reflection tomography is to transmit a narrow-bandwidth signal from a large plane wave transducer while rotating either the transducer or the object. Because of the relative motion of the object with respect to the insonifying wave, backscattered echo signals undergo Doppler frequency shifts. Spectral analysis of the echo signals allows surface integrals over planes to be measured [38] and, hence, images of reflectivity to be reconstructed. Another method involves the use of a moving transducer and the associated Doppler shifts [41]. However, the theoretical resolution of these methods is inherently less than that possible with other reflection tomography techniques [38].

Reflection tomography requires some fairly strong assumptions. An average value for attenuation must be assumed or measured to correct for reduction in apparent echo amplitudes at increasing depths. It also is assumed that the scattering is weak so that scatterers deep within the object are not "shadowed" by those close to the surface. Finally, the ultrasonic velocity is assumed constant or known so that the surfaces of integration are known. All these assumptions are similar to those made in the Born approximation. There are also similarities between reflection tomography and compound B-scan techniques, as described in Chapter 3 [29]. The main distinctions are that the apertures of compound B scanners do not usually enclose the object and that incoherent (and possibly nonlinear) compounding [42] rather than computerized reconstruction is used in a compound B scan. The bandwidths obtainable with practical transducers make it unlikely that the optimal pulse mentioned previously can be physically realized. Pulses are usually several cycles long and so significant speckle, or destructive interference, effects can occur [20, 43]. It follows that reflectivity projections are not simple integrals of scalar scattering strengths. The major problem with all coherent imaging methods in medical ultrasonics is that the tissue being imaged cannot move more than about $\lambda/4$ or the method will defocus. It is not known to what extent these assumptions affect the usefulness of reflection tomography.

8.2 Diffraction Tomography

CT techniques based on the ray approximation ignore diffraction. The ray approximation requires that the wavelength of the sound be much shorter than the scale of the inhomogeneities. Because the attenuation of soft tissues is about 0.7 dB/MHz cm^{-1} or 0.1 dB/λ and a loss of some 80 to 120 dB can be detected in the signal, 800 to 1,200 wavelengths can be penetrated in transmission. In reflection, the problem is much different. The reflectivity of tissue is only about −50 dB, so the available two-way depth of penetration is only about 150 to 350 wavelengths at best. This limits the wavelengths to no smaller than 1.3 to 0.4

mm for a 20–cm depth. Because structure on a scale much smaller than this exists in tissues, the basic ray assumption may well be violated. Significant ultrasonic diffraction effects have been demonstrated in the laboratory [44]. A method of imaging called diffraction tomography is designed to compensate for diffraction by inverting an approximate wave equation rather than using the straight-ray approximation.

Diffraction tomography reconstruction methods were described in Chapter 3. A single two-dimensional scattering experiment, using a narrow temporal bandwidth plane insonifying wave, allows estimation of the spatial Fourier transform of the object over part or all of a circle in the Fourier domain from measurement of the amplitude and phase of the scattered waves [45]. The Fourier transform domain is "filled in" by taking multiple views from many different angles of insonation, so that the circle pivots about the origin of Fourier space (Chapter 3).

As explained previously, part of Fourier space can be filled in by altering the frequency of the wave while holding the angle of view fixed [46]. The image is then calculated by a single two-dimensional inverse Fourier transform. Early two-dimensional diffraction tomography experiments used this approach [37, 46, 47]. The Fourier domain data that result from the measurements must be interpolated onto a rectangular grid to allow the fast Fourier transform algorithm to be used. Accurate interpolation schemes are required to avoid reconstruction errors [48]. Reconstruction of these data also can be performed in the image domain, similar to conventional straight-line convolution/backprojection except that the projections are "backpropagated" [49]. Inversion algorithms for spherical geometries and non-plane wave insonation also have been developed [50, 51]. Although three-dimensional reconstruction theory is well developed [49], all experimental work has so far been performed for two-dimensional geometries [37, 46–48, 51].

As previously discussed, diffraction tomography involves inverting an approximation to the true wave equation. This type of approximation to the wave equation used to derive the diffraction tomography equations is important. The inadequacy of the Born approximation for describing propagation through tissues suggests that reconstructions based on the Born approximation are unlikely to be satisfactory. The Rytov approximation is more raylike in nature and accounts, at least partially, for refraction. Theoretical and experimental studies [52] indicate the superiority of the Rytov approximation for diffraction tomography for transmission imaging.

The Rytov approximation requires the true phase (i.e., narrow-band propagation delay) of the scattered field, not just its principal value, as the data for the algorithm. However, propagation through biologic tissues usually causes phase shifts of many cycles at frequencies high enough to give adequate resolution; therefore, some sort of "phase-unwrapping" [53] procedure is necessary [52]. Phase-unwrapping algorithms usually fail when the signal amplitude decreases below the noise level. Wide-bandwidth scattering measurements may make the phase-unwrapping problem more tractable, however [54]. Other difficulties with

diffraction tomography are related to the fact that the scanner apparatus requires
a high degree of phase stability and mechanical precision and must be very fast
for live tissue.

In Section 3.4, it was demonstrated that the highest spatial frequency
components, and therefore the best theoretical spatial resolution, are obtained
from backscatter measurements. It is possible to regard backscatter imaging as
a specialized case of diffraction tomography under the Born approximation [55].

8.3 Scattering-Class Comments

Transmission imaging produces data that allow estimates of the low-frequency
region of the Fourier spectrum of the object-scattering function. This results in
estimates of Class 0 and Class 1 scattering portions of the object. Reflection
imaging can acquire data necessary for estimating the higher frequency regions
of the Fourier space. These are regions containing information about Class 2
and Class 3 scatterers. Because coherent tomographic imaging methods require
phase coherence throughout the object for the entire scanning period, maximum
resolution will be achieved only with fast data-acquisition methods. This will
require large scanning arrays and fast-pulsing and digitization methods.

Bibliography

[1] J. F. Greenleaf, S. A. Johnson, S. L. Lee, G. T. Herman, and E. H. Wood.
     Algebraic reconstruction of spatial distributions of acoustic absorption
     within tissue from their two-dimensional acoustic projections. In P. S.
     Green, editor, *Acoustical Holography*, volume 5, pages 591–603. Plenum
     Press. New York, 1974.

[2] B. D. Steinberg. Radar imaging from a distorted array–the radio camera
     algorithm and experiments. *IEEE Transactions on Antennas and Propaga-
     tion*, 29:740–748, 1981.

[3] R. H. T. Bates and B. S. Robinson. Ultrasonic transmission speckle imaging.
     Ultrasonic Imaging, 3:378–394, October, 1981.

[4] T. Sato and S. Wadaka. Incoherent ultrasonic imaging system. *Journal of
     the Acoustical Society of America*, 58:1013–1017, 1975.

[5] R. H. T. Bates and G. R. Dunlop. Inverse scattering and tomography. In
     Ultrasonic International 1977 Proceedings, pages 104–110. IPC Science
     and Technology Press. Guildford, U.K., 1977.

[6] C. R. Mol, J. Heethaar, K. Bakker, and R. M. Heethaar. Ultrasound velocity
     tomography: an imaging method. *Journal of Biomedical Engineering*,
     3:235–238, 1981.

[7] J. F. Greenleaf. Computerized transmission tomography. In P. D. Edmonds, editor, *Methods of Experimental Physics-Ultrasound*, volume 19, pages 563–589. Academic Press. New York, 1981.

[8] J. F. Greenleaf, S. A. Johnson, R. C. Bahn, B. Rajagopalan, and S. Kenue. Introduction to computed ultrasound tomography. In J. Raviv, J. F. Greenleaf, and G. T. Herman, editors, *Computer Aided Tomography and Ultrasonics in Medicine*, pages 125–136. North-Holland Publishing Company. Amsterdam, The Netherlands, 1979.

[9] D. Hiller and H. Ermert. System analysis of ultrasound reflection mode computerized tomography. *IEEE Transactions on Sonics and Ultrasonics*, SU-31:240–250, July, 1984.

[10] G. T. Herman. *Image Reconstruction from Projections: The Fundamentals of Computerized Tomography*. Academic Press. New York, 1980.

[11] Y. Censor. Finite series-expansion reconstruction methods. *Proceedings of the IEEE*, 71:409–419, 1983.

[12] J. F. Greenleaf, S. A. Johnson, and A. H. Lent. Measurement of spatial distribution of refractive index in tissues by ultrasonic computer assisted tomography. *Ultrasound in Medicine and Biology*, 3:327–339, 1978.

[13] R. Gordon, R. Bender, and G. T. Herman. Algebraic reconstruction techniques (ART) for three-dimensional electron microscopy and x-ray photography. *Journal of Theoretical Biology*, 29:471–481, 1970.

[14] J. F. Greenleaf, S. A. Johnson, W. F. Samayoa, and F. A. Duck. Algebraic reconstruction of spatial distribution of acoustical velocities in tissues from their time-of-flight profiles. In N. Booth, editor, *Acoustical Holography*, volume 6, pages 71–90. Plenum Press. New York, 1975.

[15] M. J. Haney and W. D. O'Brien, Jr. Ultrasonic tomography for differential thermography. In E. A. Ash and C. R. Hill, editors, *Acoustical Imaging*, volume 12, pages 589–597. Plenum Press. New York, 1982.

[16] R. A. Brooks and G. Di Chiro. Principles of computer assisted tomography (CAT) in radiographic and radioisotopic imaging. *Physics in Medicine and Biology*, 21:689–732, 1976.

[17] R. M. Lewitt. Reconstruction algorithms: transform methods. *Proceedings of the IEEE*, pages 390–408, 1983.

[18] R. C. Heyser and D. H. LeCroissette. A new ultrasonic imaging system using time delay spectrometry. *Ultrasound in Medicine and Biology*, 1:119–131, 1974.

[19] R. S. Mezrich, D. H. R. Vilkomerson, and K. F. Etzhold. Measurement of ultrasonic tissue characteristics by direct and phase contrast imaging. In M. Linzer, editor, *Ultrasonic Tissue Characterization I*. National Bureau of Standards (Special publication no. 453), Washington, DC, 1975.

[20] P. N. T. Wells and M. Halliwell. Speckle in ultrasonic imaging. *Ultrasonics*, 19:225–229, 1981.

[21] C. R. Crawford and A. C. Kak. Multipath artifact corrections in ultrasonic transmission tomography. *Ultrasonic Imaging*, 4:234–266, July, 1982.

[22] G. R. Dunlop. Ultrasonic transmission imaging. *Thesis, University of Canterbury, Christchurch, New Zealand*, 1978.

[23] E. L. Carstensen, W. K. Law, N. D. McKay, and T. G. Muir. Demonstration of nonlinear acoustical effects at biomedical frequencies and intensities. *Ultrasound in Medicine and Biology*, 6:359–368, 1980.

[24] J. R. Klepper, G. H. Brandenburger, J. W. Mimbs, B. E. Sobel, and J. G. Miller. Application of phase-insensitive detection and frequency-dependent measurements to computed ultrasonic attenuation tomography. *IEEE Transactions on Biomedical Engineering*, BME-28:186–201, 1981.

[25] J. G. Miller, J. R. Klepper, G. H. Brandenburger, L. J. Busse, M. O'Donnell, and J. W. Mimbs. Reconstructive tomography based on ultrasonic attenuation. In J. Raviv, J. F. Greenleaf, and G. T. Herman, editors, *Computer Aided Tomography and Ultrasonics in Medicine*, pages 151–164. North-Holland Publishing Company. Amsterdam, The Netherlands, 1979.

[26] K. M. Pan and C. N. Liu. Tomographic reconstruction of ultrasonic attenuation with correction of refractive errors. *IBM Journal of Research and Development*, 25:71–82, 1981.

[27] P. L. Carson, T. V. Oughton, W. R. Hendee, and A. S. Ahuja. Imaging soft tissue through bone with ultrasound transmission tomography by reconstruction. *Medical Physics*, 4:301–309, 1977.

[28] K. A. Dines and A. C. Kak. Ultrasonic attenuation tomography of soft tissues. *Ultrasonic Imaging*, 1:16–33, January, 1979.

[29] P. N. T. Wells. *Biomedical Ultrasonics*. Academic Press. London, 1977.

[30] L. A. Chernov. *Wave Propagation in a Random Medium*. Dover Publications, New York, 1960.

[31] G. H. Brandenburger, J. R. Klepper, J. B. Miller, and D. L. Synder. Effects of anisotropy in the ultrasonic attenuation of tissue on computed tomography. *Ultrasonic Imaging*, 3:113–143, April, 1981.

[32] A. C. Kak and K. A. Dines. Signal processing of broadband pulsed ultrasound: measurement of attenuation of soft biological tissues. *IEEE Transactions on Biomedical Engineering*, BME-25:321–344, 1978.

[33] F. Dunn, W. K. Law, and L. A. Frizzell. Nonlinear ultrasonic propagation in biological media. *British Journal of Cancer*, 45(Suppl. 5):55–58, 1982.

[34] N. Ichida, T. Sato, and M. Linzer. Imaging the nonlinear ultrasonic parameter of a medium. *Ultrasonic Imaging*, 5:295–299, October, 1983.

[35] S. J. Norton and M. Linzer. Ultrasonic reflectivity tomography: reconstruction with circular transducer arrays. *Ultrasonic Imaging*, 1:154–184, April, 1979.

[36] S. J. Norton and M. Linzer. Ultrasonic reflectivity imaging in three dimensions: reconstruction with spherical transducer arrays. *Ultrasonic Imaging*, 1:210–231, July, 1979.

[37] M. Kaveh, R. K. Mueller, R. Rylander, T. R. Coulter, and M. Soumekh. Experimental results in ultrasonic diffraction tomography. In K. Y. Wang, editor, *Acoustical Imaging*, volume 9, pages 433–450. Plenum Press. New York, 1980.

[38] G. Wade, S. Elliott, I. Khogeer, G. Flesher, J. Eisler, D. Mensa, N. S. Ramesh, and G. Heidbreder. Acoustic echo computer tomography. In A. F. Metherell, editor, *Acoustical Imaging*, volume 8, pages 565–576. Plenum Press. New York, 1980.

[39] F. J. Hilterman. Seismic imaging. In K. Y. Wang, editor, *Acoustical Imaging*, volume 9, pages 653–679. Plenum Press. New York, 1980.

[40] M. M. Sondhi. Reconstruction of objects from their sound-diffraction patterns. *Journal of the Acoustical Society of America*, 46:1158–1164, 1969.

[41] K. Nagai and J. F. Greenleaf. Ultrasonic imaging using the Doppler effect caused by a moving transducers. *Optical Engineering*, 29:1249–1254, October, 1990.

[42] D. E. Robinson and P. C. Knight. Computer reconstruction techniques in compound scan pulse-echo imaging. *Ultrasonic Imaging*, 3:217–234, July, 1981.

[43] C. B. Burckhardt. Speckle in ultrasound b-mode scans. *IEEE Transactions on Sonics and Ultrasonics*, 1:1–6, 1978.

[44] J. F. Greenleaf, P. J. Thomas, and B. Rajagopalan. Effects of diffraction on ultrasonic computer-assisted tomography. In J. P. Powers, editor, *Acoustical Imaging*, volume 11, pages 351–363. Plenum Press. New York, 1982.

[45] K. Iwata and R. Nagata. Calculation of refractive index distribution from interferograms using Born and Rytov's approximation. *Japanese Journal of Applied Physiology*, 14(Suppl. 14, no. 1):379–383, 1975.

[46] S. K. Kenue and J. F. Greenleaf. Limited angle multifrequency diffraction tomography. *IEEE Transactions on Sonics and Ultrasonics*, SU-29:213–217, July, 1982.

[47] R. K. Mueller. Diffraction tomography. I: The wave-equation. *Ultrasonic Imaging*, 2:213–222, July, 1980.

[48] M. Kaveh, M. Soumekh, Z. Q. Lu, R. K. Mueller, and J. F. Greenleaf. Further results on diffraction tomography using Rytov's approximation. In E. A. Ash and C. R. Hill, editors, *Acoustical Imaging*, volume 12, pages 599–608. Plenum Press. New York, 1982.

[49] A. J. Devaney. A filtered backpropagation algorithm for diffraction tomography. *Ultrasonic Imaging*, 4:336–350, October, 1982.

[50] J. Ball, S. A. Johnson, and F. Stenger. Explicit inversion of the Helmholtz equation for ultra-sound insonification and spherical detection. In K. Y.

Wang, editor, *Acoustical Imaging*, volume 9, pages 451–461. Plenum Press. New York, 1980.

[51] E. Wolf. Three-dimensional structure determination of semi-transparent objects from holographic data. *Optical Communication*, 1:153–156, September/October, 1969.

[52] M. Kaveh, M. Soumekh, and R. K. Mueller. A comparison of Born and Rytov approximations in acoustic tomography. In M. Kaveh, R. K. Mueller, and J. F. Greenleaf, editors, *Acoustical Imaging*, volume 11, pages 325–335. Plenum Press, New York, 1982.

[53] J. M. Tribolet. A new phase unwrapping algorithm. *IEEE Transactions on Acoustics, Speech, and Signal Processing*, ASSP-25:170–177, 1977.

[54] J. F. Greenleaf. Computerized tomography with ultrasound. *Proceedings of the IEEE*, 71:330–337, 1983.

[55] S. J. Norton and M. Linzer. Ultrasonic reflectivity imaging in three dimensions: exact inverse scattering solutions for plane, cylindrical, and spherical apertures. *IEEE Transactions in Biomedical Engineering*, BME-28:202–220, 1981.

Index

MIX
Papier aus verantwortungsvollen Quellen
Paper from responsible sources
FSC® C105338

If you have any concerns about our products,
you can contact us on
ProductSafety@springernature.com

In case Publisher is established outside the EU,
the EU authorized representative is:
**Springer Nature Customer Service Center GmbH
Europaplatz 3, 69115 Heidelberg, Germany**

Printed by Libri Plureos GmbH
in Hamburg, Germany